GEOGRAPHY
REVIEW IN 20 MINUTES A DAY

Related Titles

Algebra Success in 20 Minutes a Day
Calculus Success in 20 Minutes a Day
Chemistry Review in 20 Minutes a Day
Critical Thinking Skills Success in 20 Minutes a Day
Geometry Success in 20 Minutes a Day
Grammar Success in 20 Minutes a Day
Practical Math Success in 20 Minutes a Day
Public Speaking Success in 20 Minutes a Day
Reading Comprehension Success in 20 Minutes a Day
Reasoning Skills Success in 20 Minutes a Day
Vocabulary and Spelling Success in 20 Minutes a Day
Write Better Essays in 20 Minutes a Day
Writing Skills Success in 20 Minutes a Day

GEOGRAPHY
REVIEW IN 20 MINUTES A DAY

NEW YORK

Copyright © 2012 LearningExpress, LLC.

All rights reserved under International and Pan American Copyright Conventions.
Published in the United States by LearningExpress, LLC, New York.

Library of Congress Cataloging-in-Publication Data
Geography review in 20 minutes a day.
p. cm.
ISBN 978-1-57685-768-7
1. Geography—Study and teaching. I. Title: Geography review in twenty minutes a day.
G73.G43 2012
910—dc23

2011034386

Printed in the United States of America

9 8 7 6 5 4 3 2 1

First Edition

ISBN 978-1-57685-768-7

For more information or to place an order, contact LearningExpress at:
2 Rector Street
26th Floor
New York, NY 10006

Or visit us at:
www.learnatest.com

CONTENTS

INTRODUCTION		vii
PRETEST		1
LESSON 1	Atmosphere and Climate	11
LESSON 2	Population, Culture, and Natural Resources	21
LESSON 3	Overview of the World—Hydrosphere	31
LESSON 4	Overview of the World—Lithosphere	37
LESSON 5	Recording Geography	43
LESSON 6	Canada and the United States—Physical Geography	51
LESSON 7	Canada and the United States—Human Geography	57
LESSON 8	Latin America—Physical Geography	69
LESSON 9	Latin America—Human Geography (Part I)	77
LESSON 10	Latin America—Human Geography (Part II)	85
LESSON 11	Russia—Physical Geography	93
LESSON 12	Russia—Human Geography	99
LESSON 13	Europe—Physical Geography	107
LESSON 14	Europe—Human Geography	115
LESSON 15	Africa—Physical Geography	127

LESSON 16	Africa—Human Geography (Part I)	135
LESSON 17	Africa—Human Geography (Part II)	145
LESSON 18	East Asia—Physical Geography	153
LESSON 19	East Asia—Human Geography (Part I)	161
LESSON 20	East Asia—Human Geography (Part II)	171
LESSON 21	Southeast Asia—Physical Geography	179
LESSON 22	Southeast Asia—Human Geography	187
LESSON 23	Australia and New Zealand, Oceania, and Antarctica—Physical Geography	199
LESSON 24	Australia and New Zealand, Oceania, and Antarctica—Human Geography	207
LESSON 25	Subcontinents (Part I)	219
LESSON 26	Subcontinents (Part II)	229
LESSON 27	Subcontinents (Part III)	239
POSTTEST		253
GLOSSARY		263

INTRODUCTION

This book is an overview of the world. Each lesson is designed to be digested in about 20 minutes a day. Keep the following ideas in mind as you read each lesson.

How to Use This Book

First, take the pretest to see how much you know. Then, read each lesson. Key glossary terms are indicated in **bold**. Once you have read all the lessons and completed all the practice questions, complete the posttest. Then, complete the online test to reinforce what you learned.

Using Geography

Geography helps us understand the relationship among people, places, and environments over time. Geographers examine patterns of human and physical geography using direct observation, mapping, interviewing, statistics, and technology to describe and organize information about Earth. Geography helps you understand the past and prepare for the future.

The World in Spatial Terms

Geographers see the location of a place on Earth and where it is in relation to other places. Geographers then study how people, places, and environments are distributed across Earth.

Places and Regions

Places can be described by unique physical and human characteristics. Places reflect the relationship between humans and the physical environment. Culture influences people's perceptions of places and regions. Geographers organize Earth into regions sharing common characteristics.

Physical Systems

Physical processes (e.g., hurricanes and earthquakes) shape Earth's surface. Plants and animals depend on one another and their surroundings for survival. Understanding the characteristics and distribution of ecosystems help people address environmental issues.

Human Systems

Goods, ideas, and people connect and diverge by moving across the environment. People settle in some places but not in others for various physical and cultural reasons. The characteristics and distribution of human populations affect physical and human systems. The characteristics and distribution of cultures and technologies also influence human and physical systems. Patterns of economic interdependence vary among the world's cultures, and certain processes, patterns, and functions help determine where people settle. Cooperation and conflict among people influence the division and control of Earth's surface.

Environment and Society

Geographers study how people use their environment and how their actions affect the environment. Human actions such as consumption and conservation modify the physical environment. The physical environment affects how people act and their activities. Changes occur in the use, distribution, and importance of natural resources.

Explore World Geography

With so many outstanding Internet search engines, exciting multimedia websites, cable television, and computer programs available to assist us, our goal in geographic study and fieldwork remains exploring, mapping, and sharing the mysteries and diversity of our amazing world.

Enjoy the odyssey of geography!

GEOGRAPHY
REVIEW IN
20 MINUTES A DAY

PRETEST

Before you begin the first lesson, you may want to find out how much you already know and how much you need to learn. If that's the case, take the pretest in this chapter, which includes 50 multiple-choice questions covering the topics in this book. While 50 questions can't cover every geography concept taught in this book, your performance on the pretest will give you a good indication of your strengths and weaknesses.

Take as much time as you need to complete the pretest. When you are finished, check your answers in the answer key at the end of the chapter.

1.	ⓐ	ⓑ	ⓒ	ⓓ
2.	ⓐ	ⓑ	ⓒ	ⓓ
3.	ⓐ	ⓑ	ⓒ	ⓓ
4.	ⓐ	ⓑ	ⓒ	ⓓ
5.	ⓐ	ⓑ	ⓒ	ⓓ
6.	ⓐ	ⓑ	ⓒ	ⓓ
7.	ⓐ	ⓑ	ⓒ	ⓓ
8.	ⓐ	ⓑ	ⓒ	ⓓ
9.	ⓐ	ⓑ	ⓒ	ⓓ
10.	ⓐ	ⓑ	ⓒ	ⓓ
11.	ⓐ	ⓑ	ⓒ	ⓓ
12.	ⓐ	ⓑ	ⓒ	ⓓ
13.	ⓐ	ⓑ	ⓒ	ⓓ
14.	ⓐ	ⓑ	ⓒ	ⓓ
15.	ⓐ	ⓑ	ⓒ	ⓓ
16.	ⓐ	ⓑ	ⓒ	ⓓ
17.	ⓐ	ⓑ	ⓒ	ⓓ
18.	ⓐ	ⓑ	ⓒ	ⓓ
19.	ⓐ	ⓑ	ⓒ	ⓓ
20.	ⓐ	ⓑ	ⓒ	ⓓ
21.	ⓐ	ⓑ	ⓒ	ⓓ
22.	ⓐ	ⓑ	ⓒ	ⓓ
23.	ⓐ	ⓑ	ⓒ	ⓓ
24.	ⓐ	ⓑ	ⓒ	ⓓ
25.	ⓐ	ⓑ	ⓒ	ⓓ
26.	ⓐ	ⓑ	ⓒ	ⓓ
27.	ⓐ	ⓑ	ⓒ	ⓓ
28.	ⓐ	ⓑ	ⓒ	ⓓ
29.	ⓐ	ⓑ	ⓒ	ⓓ
30.	ⓐ	ⓑ	ⓒ	ⓓ
31.	ⓐ	ⓑ	ⓒ	ⓓ
32.	ⓐ	ⓑ	ⓒ	ⓓ
33.	ⓐ	ⓑ	ⓒ	ⓓ
34.	ⓐ	ⓑ	ⓒ	ⓓ
35.	ⓐ	ⓑ	ⓒ	ⓓ
36.	ⓐ	ⓑ	ⓒ	ⓓ
37.	ⓐ	ⓑ	ⓒ	ⓓ
38.	ⓐ	ⓑ	ⓒ	ⓓ
39.	ⓐ	ⓑ	ⓒ	ⓓ
40.	ⓐ	ⓑ	ⓒ	ⓓ
41.	ⓐ	ⓑ	ⓒ	ⓓ
42.	ⓐ	ⓑ	ⓒ	ⓓ
43.	ⓐ	ⓑ	ⓒ	ⓓ
44.	ⓐ	ⓑ	ⓒ	ⓓ
45.	ⓐ	ⓑ	ⓒ	ⓓ
46.	ⓐ	ⓑ	ⓒ	ⓓ
47.	ⓐ	ⓑ	ⓒ	ⓓ
48.	ⓐ	ⓑ	ⓒ	ⓓ
49.	ⓐ	ⓑ	ⓒ	ⓓ
50.	ⓐ	ⓑ	ⓒ	ⓓ

Directions: Select the letter of the best answer and circle it.

1. The ___________ is a layer of gases above the planet's surface.
 a. biosphere
 b. lithosphere
 c. topography
 d. atmosphere

2. ________ is/are opposite in the Northern and Southern Hemispheres.
 a. The tilt of Earth
 b. Earth's orbit
 c. The seasons of the year
 d. Earth's axis

3. __________ include(s) the atmosphere, oceans, freshwater systems, geological formations, and soils of Earth.
 a. Habitat
 b. The biosphere
 c. Biodiversity
 d. Mineral resources

4. _________ geography deals with Earth's physical features like climate, land, air, water, plants, and animal life and their relationship to each other and humans.
 a. Physical
 b. Cultural
 c. Human
 d. Political

5. ___________ is the study of population.
 a. Hydrology
 b. Demography
 c. Climatology
 d. Technology

6. Oceans, rivers, lakes, and other bodies of water make up the water-based _________.
 a. water cycle
 b. evaporation
 c. hydrosphere
 d. run-off

7. International planners use the term ____________ to describe where freshwater needs are greatest.
 a. precipitation
 b. water stress
 c. urban sewage
 d. condensation

8. The ____________ is surface land including the continents and ocean floors.
 a. lithosphere
 b. atmosphere
 c. biosphere
 d. hydrosphere

9. Plate ____________ created Earth's largest features, like continents, oceans, and mountain ranges.
 a. pollution
 b. irrigation
 c. flooding
 d. tectonics

10. Water ______________ wear(s) away soil and rock.
 a. desalination
 b. erosion
 c. valleys
 d. cycles

11. __________ protect(s) soil from wind erosion.
a. Urbanization
b. Deserts
c. Glaciers
d. Plants

12. __________, including aerial photography and satellite imaging, is used to know more about mineral deposits, and freshwater sources, as well as to survey human activities.
a. Cartography
b. Remote sensing
c. Direct observation
d. A key

13. The __________ is 0° latitude.
a. equator
b. prime meridian
c. North Pole
d. absolute location

14. Most ancient civilizations began in __________.
a. mountains
b. river valleys
c. deserts
d. plateaus

15. The raising and grazing of livestock, or __________, is a way of life for people living on the steppes.
a. agriculture
b. aquaculture
c. pastoralism
d. terracing

16. Arab expansion brought __________ to North Africa, Southwest Asia, and Central Asia.
a. Hinduism
b. Buddhism
c. Judaism
d. Islam

17. In the 1980s, the U.S. Central Intelligence Agency (CIA) financed Osama bin Laden's fight against the Russians in the mountains of __________.
a. Iraq
b. Afghanistan
c. Iran
d. Pakistan

18. __________ is a means of managing natural resources in ways that do not deplete them or cause more damage to ecosystems.
a. Sustainable development
b. The Green Revolution
c. Outsourcing
d. Qaqortoq

19. The __________, or Continental Divide, determine(s) the flow of rivers in North America.
a. Canadian Shield
b. Great Basin
c. Appalachian Mountains
d. Rockies

20. The plains of the pampas are used for __________.
a. growing cereal grains
b. cattle ranching
c. growing ornamental plants
d. training cowboys

21. __________ is the primary language of Latin America.
a. Latin
b. English
c. French
d. Spanish

22. Giant agricultural estates from the colonial era now disappearing in Latin America are called __________.

a. maquiladoras
b. caudillos
c. latifundia
d. campesinos

23. Most people in Russia live ______________.

a. west of the Ural Mountains
b. in Siberia
c. in the Caucasus region
d. in the Caspian Sea republics

24. Since 1991, the Commonwealth of Independent States (CIS) has expanded its __________ economy.

a. command
b. market
c. collectivized agriculture
d. environmental degradation and

25. Western European industry developed out of the mineral and soil resources of the ____________________.

a. Emerald Isle
b. Polders
c. North European Plain
d. Alps

26. Natural vegetation of ________ latitudes includes deciduous and coniferous trees.

a. very high
b. middle
c. very low
d. all

27. The Romance languages of Italian, Spanish, and Portuguese are Indo-European languages derived from _________.

a. Basque
b. Bantu
c. Latin
d. German

28. Overfarming, removing too much vegetation and overgrazing livestock has led to ___________ in Europe.

a. soil erosion
b. deforestation
c. increased vegetation, including trees,
d. soil erosion and deforestation

29. Fossil fuel automobile exhaust causes ______________.

a. acid rain, property damage, and environmental destruction
b. biofuel
c. natural gas
d. El Niño

30. A __________ is a triangular section of land formed by sand and silt carried downriver.

a. delta
b. cataract
c. harmattan
d. tornado

31. In Africa, ______________ climate zones can be found near the equator.

a. tropical
b. highland
c. steppe
d. altiplano

32. Darfur, site of a modern refugee crisis, is located in ____________.
a. Congo
b. Rwanda
c. Sudan
d. Cambodia

33. In the colonial era, Europeans built ________________ in Africa, Asia, and Latin America to acquire wealth from natural resources.
a. commercial plantations
b. railroads
c. port facilities
d. commercial plantations, railroads, and port facilities

34. Tanzania's Serengeti National Park, Kenya's Masai Mara, and Ghana's Kakum National Park are all ____________.
a. game reserves
b. poaching parks
c. a and b
d. none of the above

35. An undersea ____________ generates a tsunami, a giant wave that gets higher as it reaches the coast.
a. tornado
b. earthquake
c. typhoon
d. archipelago

36. Most of the tens of thousands of rivers in ____________ start in Tibet and empty into the Pacific Ocean.
a. China
b. Nepal
c. Japan
d. the Philippines

37. A hurricane in the Atlantic Ocean is a(n) ____________ in the Pacific Ocean.
a. avalanche
b. tsunami
c. typhoon
d. shogun

38. The Dalai Lama is ____________ Buddhist spiritual leader.
a. Nepal's
b. Japan's
c. China's
d. Tibet's

39. Due to ____________ control over natural resources in the Pacific, Japan attempted in the first half of the twentieth century to take military control of the region.
a. Western
b. Chinese
c. Arab
d. East Indian

40. Chinese and Korean governments have been most influenced by ____________.
a. Taoism
b. Buddhism
c. Confucianism
d. Shintoism

41. Which of the following is true about China?
a. It is heavily polluting the environment by rapidly burning fossil fuels.
b. It is disposing of large amounts of cancerous industrial waste.
c. It is a member of the World Trade Organization (WTO).
d. All of the above.

42. The Philippines, Indonesia, and Malaysia are all ______________.

a. landlocked in Southeast Asia
b. part of mainland Southeast Asia
c. archipelagoes in Southeast Asia
d. controlled by China

43. The ___________ can only be found in Southeast Asia.

a. rhinoceros
b. minke whale
c. silkworm
d. Komodo dragon

44. The ________ first settled in Cambodia and Vietnam.

a. Mons
b. Burmans
c. Khmers
d. Thais

45. The ___________ is a Chinese form of architecture also found in Southeast Asia.

a. wat
b. pagoda
c. kabuki
d. kami

46. ____________ is the largest Muslim country in the world and a site of terrorist attacks.

a. Iraq
b. Afghanistan
c. The Philippines
d. Indonesia

47. On December 26, 2004, an earthquake in the Indian Ocean created a ___________ that left 225,000 people dead and millions homeless.

a. typhoon
b. tsunami
c. hurricane
d. tornado

48. ____________ is the only nation in world that is a country and continent.

a. Antarctica
b. New Zealand
c. Iceland
d. Australia

49. In _____________, the lowest yearly temperature is about −126°F.

a. the Arctic Circle
b. East Antarctica
c. West Antarctica
d. Oceania

50. Which of the following is rich in mineral resources and grasslands for livestock grazing?

a. Australia
b. New Zealand
c. Antarctica
d. Australia, Antarctica, and New Zealand

ANSWERS

1. d.
2. c.
3. b.
4. a.
5. b.
6. c.
7. b.
8. a.
9. d.
10. b.
11. d.
12. b.
13. a.
14. b.
15. c.
16. d.
17. b.
18. a.
19. d.
20. c.
21. d.
22. c.
23. a.
24. b.
25. c.
26. b.
27. c.
28. d.
29. a.
30. a.
31. a.
32. c.
33. d.
34. a.
35. b.
36. a.
37. c.
38. d.
39. a.
40. c.
41. d.
42. c.
43. d.
44. c.
45. b.
46. d.
47. b.
48. d.
49. d.
50. a.

LESSON

1 ATMOSPHERE AND CLIMATE

LESSON SUMMARY

In this first lesson, we focus on geographic physical systems. We examine Earth's position in relation to the sun, and how that relationship affects day and night temperatures, the seasons, and people's lives. We look at how the factors of latitude, elevation, wind patterns, and ocean currents affect climate. We then survey how geographers classify Earth's climate and vegetation into **regions** with common characteristics. Finally, we examine human affects on climate change.

It's up to people like us, all of us, to address and talk about things like runaway global warming and how we can use things like remote viewing to save our planet.

—Jim Sullivan

Earth, the Sun, and the Planets

Earth and the objects revolving around it are pulled in by the sun's gravity and enormous mass. The sun is a star made of burning gases, and planets are nearly round bodies that revolve around stars without colliding with or scooping up other objects. Our solar system has eight major planets with their own unique orbits that revolve around the sun. Mercury, Venus, Earth, and Mars are the inner planets nearest the sun. Farthest from the sun are the outer planets of Jupiter, Saturn, Uranus, and Neptune. Tiny Pluto, once considered a planet, has been downgraded by the astronomical community to dwarf planet because it's more similar to the icy bodies at the edge of the solar system than the planets.

DID YOU KNOW?

Planet Earth is around 93 million miles away from the sun. Because it has an elliptical orbit, it varies between 94.5 million miles away at its furthest position and 91 million miles when it is closest.

The planets vary in size. Jupiter is largest, and Mercury is smallest. Earth is fourth in size. All the planets except Mercury and Venus have moons or other natural satellites revolving around them. Saturn has at least 61 moons, while Earth has one. (The number of moons a planet has constantly changes as astronomers make new discoveries.) The two types of planets are terrestrial and gas giants. Mercury, Venus, Earth, and Mars are terrestrial since they have solid, rocky crusts. Mercury and Venus are scorching hot, and Mars is a cold and barren desert.

DID YOU KNOW?

In our solar system, only Earth has liquid surface water and supports a variety of life forms.

The gas giant planets Jupiter, Saturn, Uranus, and Neptune are larger, more gaseous, and less dense than the terrestrial planets. They are similar to solar systems in that they have many orbiting moons and thin, encircling rings made of ice and rock. Saturn's rings can be viewed from Earth by telescope.

Asteroids, Comets, and Meteoroids

Small objects like asteroids, comets, and meteoroids revolve around the sun. There is a large asteroid belt between Mars and Jupiter. Asteroids are small, planetlike objects, some of which can cross Earth's path. Comets, made of icy dust and frozen gases, look like bright balls with long tails as their orbits take them by the hot sun. They can come from any direction and angle. Meteoroids are called meteors when they enter Earth's atmosphere. Meteorites, large meteors that survive the intense heat of atmospheric friction, can crash to earth.

DID YOU KNOW?

In 1908, a meteorite exploded just above Earth's surface near the Tunguska River in Siberia. It incinerated a forest and hundreds of reindeer, and left an eerie orange glow in the sky for months.

Atmosphere

The air we breathe is part of Earth's atmosphere. The **atmosphere**, a layer of gases above the planet's surface, is made up of 78% nitrogen and 21% oxygen, and trace elements of nine more gases. The atmosphere extends from the surface of Earth to about 372 miles above its surface.

Climate and Weather

Weather is a short-term aspect of climate indicating atmospheric conditions in one place in a limited time frame. **Climate** describes an area's typical weather patterns over a period of time. The relationship between Earth and the sun affects climate, which influences life.

DID YOU KNOW?

Seattle, Washington, has a rainy, wet climate while Santa Fe, New Mexico, has a dry, desert climate.

Practice

Fill in the blank with the correct word from the information in the preceding paragraphs.

1. ___________ is 93 million miles from the sun and is the only water-based planet to sustain a variety of life forms.

2. ___________ planets have solid, rocky crusts and are closest to the sun.

3. ___________ giant planets are furthest from the sun.

4. The ___________ starts at Earth's surface and extends 372 miles into the air.

5. ___________ is a short-term aspect of climate.

Earth's Tilt and Rotation

Because Earth's **axis** is tilted at an angle of 23.5°, not all places receive the same amount of sunlight. This tilt affects **temperature**. Direct sunlight means higher temperatures in degrees Fahrenheit (°F) or degrees Celsius (°C). In this book, we use the standard American measure of temperature, Fahrenheit.

Earth rotates on its axis, making a complete cycle in approximately 24 hours. Rotating west to east, Earth first turns the **Western Hemisphere**, then the **Eastern Hemisphere**, to the sun. Thus, one of these **hemispheres** experiences day while the other experiences night.

Earth's Revolution

While rotating on its axis, Earth also orbits around the sun. Earth completes one elliptical **revolution**, or orbit, around the sun in approximately 365 days. Revolution and tilt change the angle and amount of sun reaching a particular place. These changes follow a regular progression called seasons. As the seasons change, people in different locations experience differences in the length of days and in daily temperatures.

Seasons are opposite in the Northern and Southern Hemispheres. Spring in the Northern Hemisphere means autumn in the Southern Hemisphere.

DID YOU KNOW?

An **equinox**, when the sun's rays fall directly on the equator and daytime and nighttime hours are equal in length everywhere, happens twice a year—once in March and once in September.

The Poles

For six months a year, one pole is tilted toward the sun while the other is tilted away. The sun never sets on the North Pole from March 20 to September 23. Conversely, the South Pole experiences continuous daylight from September 23 to March 20.

Greenhouse Effect

Most of the sun's radiation is reflected back into space. The atmosphere traps some of the sun's energy and converts it into infrared radiation, or heat, which helps plants grow. This is called the **greenhouse effect**. Clouds and greenhouse gases, or atmospheric components like water vapor and carbon dioxide

(CO_2), absorb heat reflected by Earth and radiate part of it back into the atmosphere.

In recent decades a rise in atmospheric CO_2 has coincided with a rise in global temperatures. Many scientists believe this **global warming** is caused by human activities like burning coal, oil, and natural gas. These **fossil fuels** release carbon dioxide, a greenhouse gas, into the atmosphere, trapping more heat. Some scientists say global warming is a natural cycle, not a human-induced phenomenon.

Global warming will probably make weather patterns more extreme. Water will evaporate more quickly from the oceans, increasing humidity and rainfall. Rapid water evaporation from soil will dry out farm lands much more quickly.

Practice

Insert T or F to indicate whether the statement is true or false, based on what you've read.

6. ____ The tilt of Earth's axis causes six months of sun at one Pole at a time.

7. ____ Earth rotates west to east on its axis.

8. ____ Earth's revolution around the sun creates seasons, and seasons are opposite in the Northern and Southern Hemispheres.

9. ____ Day and night are always at the same time in the Eastern and Western Hemispheres.

10. ____ Global temperatures have recently increased.

Factors Affecting Climate

During Earth's annual orbit around the sun, the sun's rays fall into a pattern. Elevation and latitude determine the effects of the angles of the sun's rays on Earth. Such patterns, zones, and latitudes define climate regions. Lesson 5 goes into detail about the role of latitude and longitude in geography.

Elevation

Earth's atmosphere thins as altitude, or **elevation**, increases. Thinner air is less dense and retains less heat. Even places elevated near the equator lose temperature. There is a loss of about 3.5°F for each 1,000 feet. Thinner atmospheres filter fewer rays of the sun; it is therefore very bright but cold and snowy in the mountains.

Low Latitudes

The low latitudes lie between 30° S and 30° N. They include the Tropic of Capricorn, the **equator**, and the Tropic of Cancer. Some of the low latitudes receive the direct rays of the sun year round. Places in the low latitudes have warm or hot climates.

Tropics of Cancer and Capricorn

The Tropic of Cancer, at the latitude of 23.5 ° N, is the northernmost point at which the sun can be seen directly overhead at noon. The June 21 **solstice** marks the beginning of summer in the Northern Hemisphere. Around September 23, as Earth continues its revolution around the sun, the sun's direct rays strike the equator to mark the seasonal equinox (autumn in the Northern Hemisphere and spring in the Southern Hemisphere). Around December 22, the sun's rays directly strike the Tropic of Capricorn, at the latitude of 23.5° S. This solstice marks the beginning of winter in the north and the start of summer in the south.

Midlatitudes

The weather varies most at the midlatitudes between 30° N and 60° N in the Northern Hemisphere and 30° S and 60° S in the Southern Hemisphere. The midlatitudes have a temperate climate, not too hot or cold, but with dramatic seasonal changes. Hot and cold air masses affect the midlatitudes all year. Cold air can bring summer relief and harsh winters. Warm, wet, tropical air brings rain and snow in the winter.

High Latitudes

The North Pole, 60° N to 90° N, and South Pole, 60° S to 90° S, make up the high latitudes. The North Pole, north of the Arctic Circle (66.5° N), receives continuous daylight or twilight from about March 21 to September 23 when the Northern Hemisphere is tilted toward the sun. The Antarctic Circle (66.5° S) experiences daylight or twilight the other six months of the year.

Winds and Ocean Currents

Wind and water combine with the effects of the sun to influence Earth's weather and climate. Air moving across Earth's surface is wind. As sunlight heats Earth unevenly, rising warm air creates low pressure areas. Sinking cool air creates high pressure areas. Cool air flows in to replace warm rising air. These movements of air cause wind, which distributes the sun's energy across Earth.

Wind Patterns

Winds blow due to temperature differences across Earth. Polar air moves toward the equator, while tropical air moves toward the Poles. There are constant patterns of prevailing winds. Direction is determined by latitude and Earth's movement. Because Earth rotates west to east, global winds are displaced clockwise in the Northern Hemisphere and counterclockwise in the Southern Hemisphere. This is the **Coriolis effect,** named for the French scientist Gustave-Gaspard Coriolis.

The trade winds, named by early sailors, blow along the low latitudes. They blow northeast toward the equator from about latitude 30° S and southeast toward the equator from about latitude 30° N. Westerlies prevail in the midlatitudes, blowing west to east between 30° N and 60° N and between 30° S and 60° S. In the high altitudes, the polar easterlies blow diagonally east to west, pushing cold air toward the midlatitudes.

Ocean Currents

Cold and warm streams of water move through the oceans. **Currents** are affected by Earth's rotation, changes in air pressure, and differences in water temperature.

DID YOU KNOW?

The Coriolis effect occurs in ocean currents as well.

Horse Latitudes

When the trade winds from the north and south come together at the equator, the direct sunlight heats the air, making it rise. The result is a narrow windless band known as the **doldrums**, where there is almost no surface wind and frequent rainstorms from the moist, warm air. Two additional narrow bands of calm air, the horse latitudes, encircle the globe just north of the Tropic of Cancer and just south of the Tropic of Capricorn.

DID YOU KNOW?

Without moving air to power ships, sailors were afraid of being stranded in the hot, still weather only to lose food and other perishable cargo while their ships sat motionless. Livestock and supplies for colonies were thrown overboard to lighten the ship to escape the calm waters of the horse latitudes.

Weather and the Water Cycle

Wind and water work together to affect weather. Powered by temperature, condensation creates **precipitation.** Water vapor forms in the atmosphere from evaporated surface water. Rising air cools when evaporation increases. The water condenses into liquid droplets, forming clouds. Additional cooling brings rainfall, which lowers temperatures on a steamy day.

El Niño

Climate is the result of recurring events that alter the weather. One of these events is El Niño, a periodic warming of the ocean's surface in the eastern tropical Pacific, which causes changes in ocean currents, water temperatures, and weather in the mid-Pacific region. This phenomenon has occurred more often in the last 25 years.

DID YOU KNOW?

El Niño, Spanish for "the Christ Child," was so named because it happens around Christmas in late December. However, it does not happen every year.

In an El Niño year, the regularly low atmospheric pressure over the western Pacific rises and the normally high pressure over the eastern Pacific drops. The reversal causes trade winds to diminish or reverse direction. The change in wind pattern reverses equatorial ocean currents, pulling warm water from near Indonesia east to Ecuador to spread along the Peruvian and Chilean coasts.

Changing air pressure from El Niño influences climates around the world. Precipitation increases along the coasts of North and South America, warming waters and increasing flood risk. In Southeast Asia and Australia, drought and massive forest fires occur.

There may be a link between global warming and El Niño.

Landforms and Climate

Landforms and bodies of water influence Earth's climate patterns. Large water bodies do not heat or cool as fast as land, so water temperatures are more uniform and constant than land temperatures. Coastal lands receive the moderating influence of water bodies and therefore have less changeable weather than inland locations.

Mountain ranges affect climate and precipitation. Winds push up when they meet a mountain. The rising air cools and releases moisture (precipitation) on the side facing the wind, or the windward side. The winds then warm and dry as they move down the opposite, or leeward, side of the mountain. This causes a **rain shadow effect**, the formation of dry areas or deserts on the leeward side.

Practice

Insert T or F to indicate whether the statement is true or false, based on what you've read.

11. ___________ latitudes are found at the equator and in the tropics.

12. The windless ___________ result from warm air that rises near the equator.

13. ___________ winds help sailors along the low latitudes.

14. The ___________ latitudes include the Poles.

15. Rain shadows form on the ___________ side of mountains.

Climate Regions

Over the centuries, many have tried to classify Earth into regions with similar climates. The most frequently used classification system today is the Köppen system, first devised by German climatolgist Wladimir Köppen in 1884. He differentiated five main regions: tropical, dry, temperate (or midlatitude), severe midlatitude, and polar. A sixth, highland, was added later. Regions are subdivided into smaller ones based on soils and **natural vegetation**, the plant life growing in an area where the natural environment is unchanged by human activity.

Tropical Climates

Tropical climates are found in the low latitudes, or tropics. Tropical wet climates are hot and wet all year with a temperature over 80°F. The warm, moisture-saturated air produces rain virtually every day. The heavy rainfall, over 80 inches a year, draws out, or leaches, nutrients from soil. Wildlife is abundant.

Tropical rain forest vegetation grows in thick layers, with tall tree canopies towering over other trees and bushes. Shade plants thrive in the shadow of trees. The world's largest rain forest is the Amazon River basin. Other parts of South America, the Caribbean, Asia, and Africa have similar climate and vegetation.

Tropical dry climates have high year-round temperatures, wet summers, and dry winters. In the winter, the ground is covered with clumps of coarse grass. There are few trees in these **savannas**, or plains, found in Africa, Central and South America, Asia, and Australia.

Dry Climates

The two types of dry climates, desert and steppe, are found in the low and midlatitudes. Deserts are dry areas with little plant life. Annual desert rainfall rarely exceeds 10 inches and temperatures vary from cool at night to hot during the day. One-third of Earth is desert.

DID YOU KNOW?

The Sahara Desert covers almost a third of North Africa.

The natural vegetation of deserts includes scrub and cacti, and plants that can handle low humidity, wide temperature ranges, and low and unreliable precipitation. In some deserts, an underground spring may support an oasis of lush vegetation. Some deserts have rocks or sand dunes and others have fertile soil capable of yielding crops using irrigation.

Dry, mostly treeless grasslands called **steppes** often border deserts. Yearly steppe rainfall is 10 to 20 inches. The world's largest steppe stretches across Eastern Europe and Western and Central Asia. Steppes also exist in North America, South America, Africa, and Australia.

Midlatitude Climates

Midlatitudes include four temperate climate regions. Their variable weather patterns and seasonal changes yield a variety of natural vegetation.

Marine west coast climates exist along coastlines from 40° N to 60° N and 40° S to 60° S. Ocean winds bring cool summers and cool, damp winters along America's Pacific coast. Abundant rainfall supports abundant coniferous trees like evergreens and deciduous trees with broad leaves that change color and drop off in winter. Mixed forests with both types of trees are common.

Lands around the Mediterranean Sea have hot, sunny summers and mild, rainy winters. Natural vegetation includes thickets of woody bushes and short trees called Mediterranean scrub. Areas with similar vegetation of the midlatitudes like southwest Australia are defined as Mediterranean.

In the southeastern United States, the humid subtropical climate brings short mild winters and rain virtually all year. Ocean wind patterns and high pressure keep humidity high. Vegetation includes prairies or inland grasslands and evergreen and deciduous tree forests.

In some midlatitude regions of the Northern Hemisphere, landforms influence climate more than ocean temperatures, wind, or precipitation. A humid continental climate does not have moderating ocean winds because they lie inland. The farther north in the humid continental climate, the longer and more severe are the snowy winters and the shorter and cooler are summers. Vegetation in humid continental regions is similar to West Coast areas with evergreens outnumbering deciduous trees in northern areas. Because of the generally mild climate and rich soil, humid continental regions are some of the world's most fertile and productive agricultural zones.

High Latitude Climates

In high latitudes, freezing temperatures prevail throughout the year due to a lack of direct sunlight. The amount and variety of vegetation are limited.

The subarctic climate region lies just south of the Arctic Circle, where winters are freezing cold and summers are short and cool. Only a thin layer of the surface thaws each summer. Below is permanently frozen subsoil known as **permafrost**. There may be needled evergreens in a short summer growing season.

Closer to the poles is **tundra**. Bitter cold and winter darkness last many months, as the sun's indirect rays over long summer days provide little warmth. The layer of thawed soil is even thinner than in the subarctic, so vegetation is limited to low bushes, very short grasses, mosses, and lichens. Trees cannot grow roots in the tundra.

Snow and ice more than two miles thick cover the surface of the ice cap regions all year. Since temperatures average below freezing, lichens are the only form of vegetation to grow here.

Highland Climates

High mountain areas, even along the equator, resemble high latitude climates because the atmosphere is so thin at high altitudes. As elevation increases, temperature cools. Natural vegetation varies by elevation.

Climate Change

Climate changes occur over time. Although the causes for climate change are unclear, evidence suggests human activity has influenced some of the changes. Scientists study temperatures, greenhouse gases, and cloud cover to learn more. One **hypothesis** is that the sun's energy output may vary over time. Another hypothesis is that volcanic dust reflects sunlight back into space, cooling the atmosphere and lowering surface temperatures.

Human activities, like burning fossil fuels, release gases that mix with water in the air to form **acid rain** and snow, which can destroy forests. Fewer forests may mean climate change. Exhaust release from factories and automobile engines is heated in the atmosphere by the sun's ultraviolet rays, forming **smog**, a haze that endangers people's health. Dams and river diversions, which are intended to supply water to dry areas, may cause new areas to flood or dry out and may affect climate over the long term.

Practice

Insert T or F to indicate whether the statement is true or false, based on what you've read.

16. ____ Steppes border ice caps.

17. ____ Oceans moderate temperature.

18. ____ Temperature decreases with increased elevation.

19. ____ Automobile exhaust endangers people's health.

20. ____ Natural vegetation is the plant life growing in an area where the natural environment is altered by human activity.

Answers

1. Earth
2. Terrestrial
3. Gas
4. Atmosphere
5. Weather
6. T
7. T
8. T
9. F
10. T
11. Low
12. Doldrums
13. Trade
14. High
15. Leeward
16. F
17. T
18. T
19. T
20. F

LESSON

2 POPULATION, CULTURE, AND NATURAL RESOURCES

LESSON SUMMARY

In this lesson, we define physical and human systems and survey human interaction with the environment, which is known as human or cultural geography. We look at how human populations change; classify the basic forms of human government, economy, and culture; and explore how humans interact with plants, animals, lands, and waters that make up Earth's biosphere.

I think the greatest challenge in environmentalism and the most rewarding challenge is trying to figure out how humans can meet their needs while protecting the environment.

—Gale Norton

Biosphere

People, animals, and plants live on or close to Earth's surface. This **biosphere** includes the atmosphere, oceans, freshwater systems, geological formations, and soils of Earth. Biodiversity explains the amazing variety of plant and animal life here on Earth.

Ecosystems

An ecosystem is a community of plants and animals, or **flora** and **fauna**, dependent on one another and their surroundings to survive. Knowing the characteristics and distribution of ecosystems helps people understand environmental issues.

Earth's land, water, and air are interconnected, as are all animals and humans. Natural ecosystems are threatened by the expansion of human communities, and scientists have demonstrated that an entire species disappears completely from Earth every 20 minutes. This era of mass extinction is now being compared to the end of the last ice age 12,000 years ago and the disappearance of the dinosaurs 65 million years ago.

DID YOU KNOW?

There are about 2 million species of plants, animals, and microorganisms. Scientists estimate there are millions more living in remote, wild areas of land and water that have not been discovered or studied.

Physical Geography

Physical geography deals with Earth's physical features like land, air, water, plants, and animal life, in terms of their relationships to each other and to humans. Physical events such as tsunamis, hurricanes, and earthquakes shape Earth's surface.

Humans modify the physical environment through the use and distribution of resources. The physical environment then affects people and their activities. Geographers examine how people use their environment, how and why people change the environment, and what consequences arise from these changes.

Human, or Cultural, Geography

Human, or cultural, geography studies human activities and their relationships to the physical environment. This includes everything from population growth and urban development to economic production and consumption.

A **culture** consists of the ways a particular group of people have learned to live, including codes of behavior, symbols, language, values, and beliefs. The movements of people, goods, and ideas help geographers understand how people cooperate or compete to change or control aspects of Earth to meet their needs. Human geography can be subdivided into subfields like historical or political geography.

Practice

Fill in the blank with the correct word from the information in the preceding paragraphs.

1. The ____________ includes the atmosphere, oceans, freshwater systems, geological formations, and soils of Earth.

2. ____________ geography focuses on population growth, urban development, and economic production and consumption.

3. ____________ geography deals with Earth's physical features like climate, land, air, water, plants, and animal life in terms of their relationships to each other and humans.

4. A(n) ____________ is a community of plants and animals dependent on one another and their surroundings to survive.

5. Humans modify the physical environment through the use and distribution of ____________.

DID YOU KNOW?

From 1000 to 1800, the world population grew slowly. The world population then more than doubled from 1800 to 1950. The world population is now growing so fast it may reach 9 billion by 2050.

Demographic Transition

Population growth varies from country to country and is influenced by cultural ideas, migration, and level of development. **Demography** is the study of population. Demographers use statistics to analyze population growth. The **birth rate** is the number of births per year per 1,000 people. The **death rate** is the number of deaths per year per 1,000 people. **Natural increase**, the difference between the birth and death rates, is the growth rate of a population. **Migration** is the movement of people from place to place.

The **demographic transition** model uses birth rates and death rates to track a country's population trends. This model reveals that birth rates and death rates declined in industrialized Western European countries during the last century. Death rates fall because of abundant and regular food supplies, improved healthcare and access to medicines, and better technology and living conditions. Birth rates decline more slowly as cultural traditions change.

Most of the world's industrialized countries have transitioned from high birth rates and death rates to low birth rates and death rates. Some countries have zero population growth, which is when birth rates and death rates are equal.

For the past 40 years, birth rates have fallen in Asia (with the exception of China), Africa, and Latin America. Birth rates are nevertheless still higher in the developing world than in the industrialized world. Families remain large in these areas due to cultural beliefs about marriage, family, and the value of children.

People in rural areas often believe having a large number of children will provide a source of labor to help them farm the land. In these developing regions, a high number of births and low death rates mean increased population growth.

Some countries have experienced negative population growth. This is when the annual death rate exceeds the annual birth rate.

Challenges of Population Growth

Rapid population growth increases the difficulty of producing enough food for everyone. On every continent except Africa, food production has risen since 1950. Africa—especially in the sub-Saharan countries—is starving due to a lack of investment in agriculture, governmental mismanagement, severe weather conditions, and warfare.

Rapidly growing populations use their resources more quickly, and there may be housing, water, and food shortages. Countries with a large population of infants and children do not have enough workers to produce food for everyone.

Population Distribution

World **population distribution** is uneven, and is influenced by migration and Earth's physical geography. Of the 30% of Earth's surface that is land, two-thirds of it, including frozen tundra, barren deserts, and high mountain peaks, remains inhospitable. Humans live on the remaining third where there is fertile soil, available water, and a livable climate. Europe and Asia are the most densely populated areas of the world—Asia contains nearly 60% of the world's population. Where population is concentrated, people tend to live in metropolitan areas, which include cities and their surrounding areas.

DID YOU KNOW?

Today most people in Europe, North America, South America, and Australia live in or near urban areas.

Population Density

Geographers determine **population density**, or how crowded a place is, by how many people live in a square mile or kilometer. Density is calculated by dividing total population by total land area. Population density varies widely by country. Singapore has one of the highest population densities, about 16,540 people per square mile. Canada has about 9 people per square mile, though its major cities are densely populated.

Countries with similarly sized populations do not have the same population densities. Both Bolivia and the Dominican Republic have populations of about 8.9 million people, but the Dominican Republic, part of a small island, has 471 people per square mile while Bolivia has only 21 people per square mile.

Population density measures do not account for uneven population distribution within a country. For example, 90% of Egypt's population lives along the Nile River, because most of the country is desert. Some geographers prefer to measure population density against hospitable land area.

Population Movement

Population generally moves from city to city, city to suburbs, and rural areas to cities. The growth of city populations by migration is called **urbanization**, which occurs primarily as rural people move to more prosperous cities in search of a better life. Although many rural populations have grown, the amount of farmland has not increased, and many rural immigrants seek to find urban jobs in manufacturing and services.

About 50% of the world's population lives in cities.

DID YOU KNOW?

From 1950 to 2005, the population of Mexico City rose from 3 million to 19 million. Cities in Latin America and Asia all grew during the same time frame. Many of these cities contain most of their country's population. About a third of Argentina's population, for example, lives in Buenos Aires.

Population movement also occurs between countries. Emigrants move from their countries of birth to a new homeland, where they are called **immigrants**. Economic pull factors have lured millions of people from Africa, Asia, and Latin America to the wealthier countries of North America, Europe, and Australia. These **refugees** are escaping disaster or persecution, and are pushed out of their homelands by war, food shortages, or other problems.

Practice

Insert T or F to indicate whether the statement is true or false, based on what you've read.

6. ____ Demography is the study of population.

7. ____ Geographers determine population density by how many people live in a square mile or kilometer.

8. ____ Humans live on 80% of Earth's land.

9. ____ The growth of city populations by migration is called urbanization.

10. ____ Population growth occurs when the annual death rate exceeds the annual birth rate.

Elements of Culture

Culture is the way of life of a group of people who share similar beliefs and customs. Culture can be understood by looking at language, religion, daily life, history, art, government, and economy. The interaction of cultures can spread new ideas, establish trading relationships, cause wars, and build political partnerships.

TIP

Geographers divide the Earth into **culture regions** defined by common elements like language and religion.

Language

Language relays information, experiences, traditions, and values. People speak in **dialects**, or variations of a language in which meanings and pronunciations vary. Languages are divided into **language families** with common roots like the Indo-European family, which dates back to the Bronze Age.

Religion

A religion is a set of beliefs and practices about ultimate reality. Each religion has its own symbols, sacred texts and sites, celebrations, and styles of worship. Religion often gives people identity, as it can dictate how people speak, eat, dress, and act. Religious symbols and stories have shaped cultural expressions in painting, architecture, and music.

Geography of Religions

Religion as a part of culture is diffused through migration, missionary work, trade, and war. Buddhism, Christianity, and Islam spread mostly through missionary work. Hinduism, Judaism, and Sikhism are associated with a culture group and people are usually born into these religions.

Indigenous Religions

Aboriginal or indigenous **animists** believe we are part of the beauty of nature. **Oral traditions** and stories, music and dance, rituals, face painting, and masks are used by natives indigenous to Africa, America, and Australia. Japanese Shintoists worship nature spirits called *kami*. There are many Native American religions, many of which worship a Great Spirit, a spirit world, hunting, and the harvest.

Culture Regions

A **culture region** is a geographical area that contains communities with traits in common. Regions may have similar social groups, forms of government, or economic systems. Their histories, religions, and art forms may have similar traditions. Internal factors like new ideas, lifestyles, and inventions change cultures over time. Spatial interaction through trade, migration, and wars are external causes of cultural change. **Cultural diffusion** occurs when new knowledge spreads from one culture to another.

The earliest humans spread their culture as they wandered to hunt and herd. These **nomads** moved from place to place seeking food, water, and grazing lands. During the last ice age, about 10,000 years ago, they settled in hills and then in river valleys and fertile plains. There they set up permanent villages, and farmers grew crops on the same land every year. This shift from gathering food to producing food through farming created an early revolution in agriculture. However, what we know today as the agricultural revolution occurred in the eighteenth and nineteenth centuries, when technological innovations created an explosion of agricultural products.

Beginning in the 1700s and 1800s, countries began to use power-driven machines and factories to mass-produce goods quickly and cheaply. This **industrialization** spurred social changes. Cities grew larger as people left farms for factory and mill jobs.

Toward the end of the twentieth century, computers made it possible to store huge amounts of information and instantly send it all over the world. This information revolution linked the cultures of the world more closely than ever.

Cultural Hearths

A **cultural hearth** is a site of innovation from which ideas, goods, and technology spread to many cultures. Early cultural centers or hearths of civilization in Egypt, Iraq, Pakistan, China, and Mexico spread ideas and practices to surrounding areas. Each of these hearths was located near a major river or water source, and farming settlements were established in areas with fertile land and mild climates. Inhabitants dug canals and ditches to irrigate the land and sell surplus crops.

Without the need to constantly produce crops, some people began to specialize in metalworking and shipbuilding. Such new technologies inspired long distance trade. Wealth from trade created governments that organized harvests, built cities, and raised armies for defense.

Cultural Diffusion

Ideas and practices are received from trade and travel through cultural diffusion. People migrate to escape wars, persecution, and famine. Mass migrations can also be forced, as during the centuries of African slave trade. Economic opportunities and religious or political freedom may also inspire migration. Migrants take their cultural ideas and practices to new lands, where they blend them with those already there, as has happened with metalworking and indigenous cultures.

Practice

Fill in the blank with the correct word from the information in the preceding paragraphs.

11. Cultural ____________ occurs when new knowledge spreads from one culture to another.

12. Indigenous ____________ believe we are all part of the beauty of nature.

13. Early cultural hearths were located near a ____________ source.

14. A culture ____________ has countries with traits in common.

15. The ____________ revolution spurred social changes as cities grew.

Resources in the World Economy

People are dependent on the world's natural resources for survival, and the growth of the **global economy** is making the people of the world more interdependent, or reliant, on each other in a process called **globalization**. Natural resources are extracted and traded around the world. Still, as important as they are to modern life, certain economic activities can threaten future human access to these resources.

Resource Management

Earth provides all the elements needed to sustain life. **Natural resources** can be used for food, fuel, or other needs. Renewable resources cannot be used up. They are replaced naturally, or are grown again right away. Wind, sun, water, and forests are renewable resources. Nonrenewable resources from Earth's crust, including minerals and fossil fuels, cannot be replaced and therefore must be conserved.

Conservation is the process by which humans manage vital resources so both present and long-term needs are met. Many countries produce **hydroelectric power**, which is generated from falling water. Solar power is produced by the sun's energy. Electricity derived from nuclear energy, or a controlled atomic reaction, is another renewable resource. However, like fossil fuel, nuclear fuel poses dangerous waste problems.

Economies and World Trade

Natural resources are not evenly distributed across Earth. Countries specialize in **economic** activity best suited to their resources. Countries with varying levels of economic development have thus become increasingly interdependent through world trade.

The global reach of **developed countries** has ignited resentment in some **developing countries**. Militant groups have struck back through terrorism, the use of violence to create fear in a population or region. These groups are small and limited in resources. They seek to heighten their influence and promote and control changes in society through fear.

Economic Development of Resources

Geographers believe there are four types of economic activities:

1. *Primary*—the taking or use of natural resources directly from Earth. This includes farming, grazing, fishing, forestry, and mining.

2. *Secondary*—the use of raw materials to produce something new and valuable. Manufacturing automobiles and assembling electronics are examples.

3. *Tertiary*—these activities provide services, like teachers and lawyers do, to people and businesses.

4. *Quaternary*—processing, managing, and distributing information in the new information economy. This includes white-collar professionals in business, education, government, and information processing and research.

DID YOU KNOW?

Industrialization, or the spread of industry, helps a country develop. Industrialized countries with technology and manufacturing, such as the United States, are called developed countries.

Farming and Industry

Farmers in developed countries work in commercial farming, raising crops and livestock to sell in the market. With the development of modern techniques, only a small percentage of farmers in developed countries grow food to support the population. Most farmers use **subsistence farming** techniques, harvesting just enough for their own family's needs.

New industrial countries move from primarily agricultural to manufacturing and industrial activities. Developing countries like Mexico and Malaysia work toward more manufacturing and technology. In the many developing countries of Africa, Asia, and Latin America, agriculture predominates.

Water and Land Pollution

Earth's natural waters are normally renewable, but the **water cycle** can be interrupted by oil spills from offshore tankers and drilling rigs. Industrial chemical waste enters and pollutes the water supply. Farm fertilizers and **pesticides** can seep into groundwater just like animal waste and untreated sewage can.

Land **pollution** includes chemical waste that poisons fertile topsoil. Solid waste dumped in landfills is another form of land pollution. Radioactive waste from nuclear power plants and toxic runoff from chemical processing plants can seep into soil, too.

Air Pollution

The burning of fossil fuels by industries and vehicles is a main source of air pollution. This process gives off poison gases that are damaging to people's health. Acidic chemicals in air pollution combine with precipitation to form acid rain. Acid rain eats away at building surfaces, kills fish, and destroys forests that provide habitats for animals, prevent soil erosion, and carry on photosynthesis, the process through which plants take in carbon dioxide and produce carbohydrates and oxygen with the presence of sunlight.

TIP

Increases in temperature cause glaciers and ice caps to melt, raising the level of the world's oceans, which could in time flood coastal cities.

Practice

Insert T or F to indicate whether the statement is true or false, based on what you've read.

16. ____ Fossil fuels such as oil and coal and other nonrenewable resources must be conserved.

17. ____ Most of the farmers in developing countries are engaged in subsistence farming, harvesting just enough for their own family needs.

18. ____ Pollution occurs when unclean or toxic elements destroy the air, water, and land.

19. ____ Forests provide habitats for animals, prevent soil erosion, and carry on photosynthesis.

20. ____ Acidic chemicals in air pollution combine with precipitation to form acid rain.

Answers

1. Biosphere
2. Human or Cultural
3. Physical
4. Ecosystem
5. Resources
6. T
7. T
8. F
9. T
10. F
11. Diffusion
12. Animists
13. Water
14. Region
15. Industrial
16. T
17. T
18. T
19. T
20. T

LESSON 3

OVERVIEW OF THE WORLD—HYDROSPHERE

LESSON SUMMARY

We look in this lesson at the importance of the water cycle. We explore saltwater bodies such as the oceans, as well as freshwater lakes, rivers, and streams. We examine how saltwater can be converted into freshwater, what groundwater is, and why it is important. We explore the scarcity of water and the destructive capabilities of floods.

More than one-half of the world's major rivers are being seriously depleted and polluted, degrading and poisoning the surrounding ecosystems, thus threatening the health and livelihood of people who depend upon them for irrigation, drinking and industrial water.

—Ismail Serageldin, Chairman, World Commission on Water for the 21st Century

Hydrosphere

About 70% of Earth's surface is water. Oceans, rivers, lakes, and other bodies of water make up the water-based **hydrosphere**. All living organisms need water to survive. The hydrosphere includes all liquid and frozen surface water, groundwater, and water vapor. About 97% of the hydrosphere is saltwater in the oceans, seas, and a few large saltwater lakes. The balance is freshwater in lakes, rivers, and groundwater.

Water Cycle

The **water cycle** is the regular movement of water from the oceans to the air to the land and back to the oceans. The sun's energy evaporates water from Earth's surfaces, changing liquid water into vapor or gas. Frozen water can be trapped in ice caps for thousands of years, or it can rest on snowy mountain peaks for months until warm seasonal weather melts it away. Water vapor rising up from the ocean, other bodies of water, and plants gather in the air. The amount of vapor the air holds depends on its temperature—thin, warm air holds more water vapor than cool, dense air.

As warm air cools it cannot hold all its water vapor. Through **condensation,** excess water vapor morphs into liquid water. These tiny water droplets form clouds. When clouds have more water than they can hold, the moisture falls to Earth as precipitation, which can be rain, sleet, or snow depending on air temperature and wind conditions.

Precipitation gathers in streams and lakes or seeps below Earth's surface as **groundwater**. Eventually, groundwater flows back to the surface, where it again evaporates along with other bodies of water, and the cycle is renewed.

DID YOU KNOW?

The amount of evaporated water from lakes, streams, and oceans is about the same amount that falls back to Earth. The amount varies very little each year. The total volume of water in the water cycle is constant.

Oceans

Ninety-seven percent of Earth's water is a huge continuous body of water circling the planet. Geographers break this water expanse into five oceans: the Pacific, Atlantic, Indian, Arctic, and Southern (or Antarctic). The Southern Ocean extends from the coast of Antarctica to 60° S latitude. The Pacific is the largest ocean. It covers more area than all the Earth's surface land combined, and at one point is so deep it could cover Mount Everest with a mile to spare.

Seas, gulfs, and bays are small saltwater bodies. They are smaller than oceans and are usually partly enclosed by land. For example, the Mediterranean, one of the world's largest seas, is almost entirely enclosed by southern Europe, North Africa, and southwestern Asia.

DID YOU KNOW?

The Gulf of Mexico is nearly encircled by the coasts of the United States and Mexico. Around the world, there are 66 separate seas, gulfs, and bays and many smaller divisions.

Saltwater to Freshwater

Saltwater from the oceans is too salty for drinking, farming, and manufacturing. Today, there is a concerted effort to try to turn ocean water into freshwater. Through **desalination**, the salt is removed to make freshwater. Only a small amount of freshwater is made through desalination because the process is expensive and consumes a lot of energy. Desalination is used, though, by countries in the Middle East and North Africa because freshwater sources are so scarce there and they have the energy supplies that enable them to run desalination plants inexpensively.

Practice

Fill in the blank with the correct word from the information in the preceding paragraphs.

1. Nearly all of the hydrosphere is ___________ water.

2. The hydrosphere includes all liquid and ___________ surface water.

3. The water ___________ is the regular movement of water.

4. The ___________ evaporates water from the surfaces of the world's oceans, lakes, and streams.

5. ___________ can be rain, sleet, or snow depending on air temperature and wind conditions.

6. Precipitation sinks into the ground and gathers in streams and as ___________.

7. ___________ percent of Earth's surface is water.

8. The ___________ is the largest and deepest ocean.

9. Today, there is a concerted effort to turn ocean water into freshwater through ___________.

10. Geographers have identified five oceans: the Pacific, Atlantic, Indian, Arctic, and ___________.

Freshwater Bodies

About 3% of Earth's water is freshwater, and most of that is not fit for human consumption. Sixty-nine percent of freshwater is contained in frozen ice caps and glaciers. Lakes, streams, and rivers contain less than 1% of Earth's freshwater. About 30% of freshwater may be found beneath Earth's surface.

Lakes, Streams, and Rivers

A lake is a body of water completely surrounded by land. Most lakes contain freshwater, but some, like Southwest Asia's Dead Sea and Utah's Great Salt Lake, are saltwater remains of ancient seas. Most lakes remain where glacial movement cut deep valleys and built up dams of rock and soil that held back melting ice water. North America has thousands of glacial lakes.

Streams and rivers form from flowing water. A spring, overflowing lake, or **meltwater** may be the source or beginning of a stream. Streams may combine to form a river. A river is a large stream of higher volume following a channel along a particular course. When rivers join together, the resulting major river systems can flow for thousands of miles. Rain, runoff, and water from tributaries (small branches of rivers) enlarge rivers as they flow on to a gulf, sea, or ocean. The mouth of the river is where the river empties into another body of water.

Groundwater

Fresh water within the earth's surface is called groundwater. Groundwater comes from rain and melted snow filtered through soil and from lake and river water that seeps into the ground. Wells and springs tap into the groundwater and are important freshwater sources for many rural people and some city dwellers. An **aquifer** is an underground porous rock layer usually saturated by very slow water flows. Aquifers and groundwater are important sources of fresh water.

Scarcity of Freshwater

International planners use the term water stress to describe where freshwater needs are greatest. Water resources are weighed against population size and growth to determine water stress. China, India, Southwest Asia, Mexico, and parts of Russia will face water shortages due to pollution of their freshwater. Untreated sewage dumped into streams and rivers causes water pollution.

DID YOU KNOW?

In developing countries, only about 25% of the water supplies are treated for contaminants.

Climate change has greatly lessened the amount of water runoff for irrigation, drinking, and industry in the United States and the world. Droughts have increased in frequency. Glaciers are receding faster. Water supplies are dwindling. Freshwater levels are going down as temperatures are increasing. Water conservation is going to become an increasingly important issue in the future.

Dangerous Floodwaters

Floods kill about 35% of the people who die from natural disasters. Population density problems cause people to migrate and settle in hazardous parts of river flood plains and deltas where they are more prone to die in floods. Flooding has increased because deforestation and urbanization do not absorb or halt runoff.

In 2004 and 2011, **tsunamis** with giant waves of over 100 feet caused by earthquakes measuring between 9.0 and 9.3 on the **Richter scale** (10 is the highest level) hit Southeast Asia and Japan, respectively. Because there was no early warning system in Indonesia and in other places along the path of the 2004 tsunami, 187,000 people were killed and 43,000 more went missing as entire villages in 13 countries were swept away. Japan lost 25,000 lives in 2011 and was able to evacuate many survivors.

In 2005, there were 28 tropical storms and 15 hurricanes. All along the coast of the Gulf of Mexico, levees (dams to stop water that run along the banks of a river or canal) were breached and countless homes were destroyed by the floodwaters created by Hurricane Katrina. Eighty percent of New Orleans was flooded. In the aftermath, many Americans criticized the federal, state, and local governments for a lack of preparedness and failure to respond in a timely, effective, and organized fashion.

Practice

Insert T or F to indicate whether the statement is true or false, based on what you've read.

11. ____ Southwest Asia's Dead Sea and Utah's Great Salt Lake are freshwater bodies.

12. ____ A river is a large stream of higher volume following a channel along a particular course.

13. ____ The mouth of the river is where the river abruptly ends.

14. ____ Aquifers and groundwater are important sources of fresh water.

15. ____ By 2025, 25% of Africa will experience water shortages or stress.

16. ____ Untreated sewage causes water pollution.

17. ____ Climate change has increased the amount of water runoff for irrigation, drinking, and industry in the United States and the world.

18. ____ Floods kill about 35% of the people who die in natural disasters.

19. ____ Tsunamis are earthquake-driven giant waves that should be detected by early warning systems.

20. ____ Many Americans criticized the federal, state, and local governments for a lack of preparedness and failure to respond to Hurricane Katrina in a timely, effective, and organized fashion.

Answers

1. Salt
2. Frozen
3. Cycle
4. Sun
5. Precipitation
6. Groundwater
7. 70
8. Pacific
9. Desalination
10. Southern
11. F
12. T
13. F
14. T
15. F
16. T
17. F
18. T
19. T
20. T

OVERVIEW OF THE WORLD—LITHOSPHERE

LESSON SUMMARY

This lesson explains the different types of landforms. It also details how internal and external processes shape Earth's surface.

Our earth is very old, an old warrior that has lived through many battles. Nevertheless, the face of it is still changing, and science sees no certain limit of time for its stately evolution. And the secret of it all—the secret of the earthquake, the secret of the "temple of fire," the secret of the ocean basin, the secret of the highland—is in the heart of the earth, forever invisible to human eyes.

—Reginald Aldworth Daly, *Our Mobile Earth* (1926)

Lithosphere

The **lithosphere** is the outermost layer of Earth. It includes the **crust** and the upper layer of the **mantle** just beneath the crust. This is true of the surfaces of continents and islands as well as land beneath oceans. Continents and islands make up about 30% of the lithosphere, which averages about 60 miles deep, though older parts of the lithosphere are thicker.

Landforms

Landforms are the natural features of Earth's surface. Landforms have a specific shape or elevation and often contain rivers, lakes, and streams. Bodies of water are considered landforms, too, as are the formations under the water's surface. Underwater landforms are diverse, just like those on dry land—the ocean floor is a flat plain in some places and in other places includes mountain ranges, cliffs, valleys, and deep trenches.

From space, the most visible landforms are the seven continents: Africa, Antarctica, Asia, Australia, Europe, North America, and South America. Australia and Antarctica stand alone. Europe and Asia make up the landmass called Eurasia, which is linked with Africa by the Sinai Peninsula in northeastern Egypt. The narrow strip called the Isthmus of Panama connects North and South America.

DID YOU KNOW?

At the Sinai Peninsula, the man-made Suez Canal separates Asia and Africa.

The **continental shelf** is an underwater extension of the coastal plain. On average, continental shelves slope out from land for 50 miles, though they can stretch out much further. For example, the Siberian shelf in the Arctic Ocean extends for 930 miles. The shelves descend gradually to a fairly uniform depth of 460 feet. At that point, a sharp drop called a shelf break marks the beginning of the continental slope. The continental slope then drops down to the ocean floor.

Earth's Depths and Heights

The heights and depths of Earth's surface contrast sharply. The highest point on Earth is Mount Everest in South Asia at 29,028 feet above sea level. The lowest point on dry land is the Dead Sea in Southwest Asia at 1,349 feet below sea level. The deepest depression on Earth is the Mariana Trench southwest of Guam in the Pacific Ocean at around 36,000 feet deep.

Practice

Fill in the blank with the correct word from the information in the preceding paragraphs.

1. The ___________ is surface land including the continental and ocean basins or land beneath the oceans.
2. The continental ____________ is an extension plain.
3. ___________ of the seven continents are independent land masses.
4. The ___________Trench is the deepest depression on Earth.
5. Mount ___________ is the highest point on Earth.

Forces of Change in Earth

The intense heat and pressure at the center of Earth, as well as the shifting of massive **tectonic plates**, drives volcanoes and earthquakes to enrich and renew its surface. These natural processes can also disrupt and destroy human life, so scientists try to predict their occurrences. Sensors in the ground, trees, and rocks in the landscape tell of past tsunamis and earthquakes.

Earth's Layers

Earth has three layers: the **core**, mantle, and crust. The very center of Earth is an iron and nickel inner core as hot as the sun and under great pressure. The liquid outer core surrounding the inner core is a band of melted iron and nickel.

The mantle is the thick layer of hot, dense rock covering the outer core. The mantle is made of silicon, aluminum, iron, magnesium, oxygen, and other elements. This mixture is always rising, cooling, sinking, and rising again.

DID YOU KNOW?

The mantle releases 80% of the heat generated by Earth's interior.

The crust is the outer layer and rocky shell of Earth's surface. The thin crust is about two miles thick under the ocean and up to 75 miles thick under mountains. The crust is broken down into more than a dozen great slabs of rock called plates. Plates float on a partially melted layer in the upper portion of the mantle. These plates carry Earth's oceans and continents.

Plate Tectonics

Plate tectonics are the physical processes that created many of Earth's physical features. Plates move very slowly—pulling, grinding, and sliding about an inch a year—and are carried by **magma** (molten rock) currents, created by heat from the core. They constantly change the face of the planet by creating volcanoes, causing earthquakes, and pushing up mountains. When plates spread apart, magma pushes up from the mantle to form ridges. When plates bump or one plate slides under another plate, a trench is formed.

Plate Movement and Continental Drift

Earth did not look the same 500 or 100 million years ago. There was first a giant supercontinent called Pangaea. Over millions of years, the continents broke away and drifted apart in a process called **continental drift**. The broken pieces then recombined in different ways, forming the continents as we know today.

DID YOU KNOW?

The face of the Earth has been changing for 2.5–4 billion years. It is still changing.

Colliding and Spreading Plates

Giant colliding plates form mountains. The Himalayan mountain range was formed by the Indian landmass drifting against Eurasia. A mountain can form from a sea plate colliding with a continental plate.

In **subduction**, the heavier sea plate dives beneath the lighter continental plate. Crashing into Earth's interior, the sea plate becomes molten material. As magma, it bursts through the crust as a volcanic mountain.

In **accretion**, massive layers of debris pile up as the sea plate slides under the continental plate. This process levels seamounts, underwater mountains with steep sides and sharp peaks, and piles up the resulting debris in trenches. Continents grow outward this way. North America began to expand into the Pacific Ocean over 200 million years ago through accretion.

New land can form from the convergence of two sea plates. One plate moves under the other and forms an island chain at the boundary.

Spreading of sea plates can create a rift or deep crack. This process allows magma within the earth to well up between plates. The magma hardens, creating undersea volcanic mountains or ridges and some islands. Spreading occurs in the middle of the floor of the Atlantic Ocean, pushing Europe and America further apart.

Practice

Insert T or F to indicate whether the statement is true or false, based on what you've read.

6. _____ The movement of plates created Earth's largest features like continents, oceans, and mountain ranges.

7. _____ The Himalayan mountain range was formed by the Indian landmass drifting against Eurasia.

8. _____ North America began to expand into the Pacific Ocean over 200 million years ago through subduction.

9. _____ The divergence of plates forms island chains.

10. _____ Earth has always remained the same physically.

Folds and Faults

Moving plates squeeze the Earth's surface until it buckles to create **folds**, or bends in layers of rock. Plates may also grind or slide past each other creating cracks in Earth's crust called **faults**. The San Andreas Fault in California is a famous example. Faulting occurs when the folded land cannot be bent further. Earth's crust cracks and breaks into huge blocks. The blocks then move along the faults in different directions, grinding against each other. The resulting tension may release small jumps or minor tremors on Earth's surface.

Earthquakes

Earthquakes are sudden violent movements along a fault line, or tremors caused by volcanic activity. The shaking changes the land surface and the ocean floor.

DID YOU KNOW?

An earthquake in Alaska in 1964 sent a piece of land 38 feet in the air.

Earthquakes often occur where plates meet. When plates stick, tension builds. The strain gets so intense that rocks snap and shift. The energy sent along the fault causes the ground to shake and tremble. The shock waves send the rock apart from the area where it snapped.

Earthquakes are common in California and Japan, because these places are situated on the Ring of Fire, one of the most earthquake-prone areas on the planet. The Ring of Fire is a volcanic and earthquake zone of activity around the perimeter of the Pacific Ocean where plates cradling the Pacific meet the plates holding the continents surrounding the Pacific.

DID YOU KNOW

North America, South America, Asia, and Australia are affected by their locations on the Ring of Fire.

Volcanic Eruptions

Volcanic mountains form when lava or magma breaks through Earth's crust. They often form along plate boundaries where one plate plunges beneath another, for example, along the Ring of Fire. The rocky plate melts as it dives downward into the hot mantle. If the molten rock is too thick, flow is blocked and pressure builds. Ash and gas clouds may spew, creating a funnel. Through the funnel, red-hot magma shoots to the surface. The lava flow may create a large cone topped by a crater, a bowl-shaped depression, at its mouth.

Volcanoes may rise far from plate boundaries. Hot spots in Earth may blast up to the surface as volcanoes. As moving plates move over hot spots, molten rock flowing from inside Earth may create volcanic island chains like the Hawaiian Islands.

DID YOU KNOW?

Molten rock may heat underground water, creating hot springs or geysers like Yellowstone National Park's Old Faithful.

Practice

Fill in the blank with the correct word from the information in the preceding paragraphs.

11. The bowl-shaped depression at the top of the volcano is its ____________.

12. The tension and tremors from colliding plates causes ____________.

13. Moving plates squeeze Earth's surface until it buckles to create ____________, or bends in layers of rock.

14. The earthquake-prone Ring of ____________ can be found throughout the perimeter of the Pacific Ocean.

15. Plates may also grind or slide past each other, creating cracks in the earth's crust called ____________.

External Forces of Change

External forces of change like **weathering** and **erosion** shape the surface of Earth.

Weathering

Weathering breaks down rocks. There are two kinds of weathering. Physical weathering is when large rock masses are physically broken down into small pieces due to atmospheric conditions. An example is when freezing water expands and causes a rock to break apart. Chemical weathering changes the chemical makeup of rocks. An example is when rainwater containing carbon dioxide from the air dissolves certain rocks like limestone.

DID YOU KNOW?

Caves are often formed by chemical weathering.

Wind Erosion

Erosion wears away Earth's surface through wind, glaciers, and moving water. Wind erosion occurs when dust, sand, and soil move from one place to another. Plants protect land from wind erosion. In dry places or where trees and plants have been cut down, winds pick up a lot of soil and blow it away. On the positive side, winds can deposit large amounts of mineral-rich soil in other places.

Glacial Erosion

Large bodies of ice called **glaciers** move across Earth's surface. Glaciers form as snow layers compress into ice. Their weight causes glaciers to move slowly downhill or spread. They pick up rocks and soil in their paths, altering the landscape. They can carve out valleys, alter rivers, destroy forests, and wear down mountains. As glaciers melt and recede they leave large piles of rock and debris known as **moraine.**

Moraines can form long land ridges or form dams that hold water and create glacial lakes.

There are two types of glaciers: sheet glaciers and mountain glaciers. Sheet glaciers, like those covering Greenland and Antarctica, are broad flat sheets of ice. They advance a few feet in winter and recede in summer. Large blocks of ice break off coastal edges and float in the ocean as icebergs. Mountain glaciers originate in cold, high mountaintop valleys and carve out U-shaped valleys as they move downhill. Glaciers that cover an entire mountain are called ice caps.

Water Erosion

Water erosion is the result of springwater and rainwater flowing downhill and wearing away soil and rock. The water forms a gully and then a V-shaped valley. Valleys may erode to form canyons like the Grand Canyon. Pounding ocean waves erode coastal cliffs, make rock into sandy beaches, and carry sand away to other coastal areas.

Practice

Insert T or F to indicate whether the statement is true or false, based on what you've read.

16. ______ Physical weathering occurs when large rock bodies break off into small pieces due to atmospheric conditions.

17. ______ Greenland and Antarctica are mountain glaciers.

18. ______ Caves are often the result of chemical weathering.

19. ______ Moraines are icebergs.

20. ______ The Grand Canyon and ocean coastlines formed from water erosion.

Answers

1. Lithosphere
2. Shelf
3. Two
4. Mariana
5. Everest
6. T
7. T
8. F
9. F
10. F
11. Crater
12. Earthquakes
13. Folds
14. Fire
15. Faults
16. T
17. F
18. T
19. F
20. T

LESSON

5 RECORDING GEOGRAPHY

LESSON SUMMARY

Geographers use a variety of specialized research tools to conduct their work. In this lesson, we examine the geographer's craft. We examine the tools and research methods geographers use, survey the careers available in geography, and examine geography's connection to other disciplines. We specifically focus on how to read a map using lines of latitude and longitude and other representative symbols.

A good plan is like a road map; it shows the final destination and usually the best way to get there.

—H. Stanley Judd

Direct Observation

Geographers visit a place to amass specific information about it and its geographic features. They use remote sensing techniques such as aerial photography and satellite imaging to learn more about mineral deposits or freshwater sources, or to survey human activities on Earth.

Mapping (Globes and Maps)

Cartography involves designing and making maps. Maps demonstrate information more graphically than writing. Cartographers convert complex information into more understandable, visual forms showing locations, features, patterns, and relationships of people, places, and things. Every type of **map projection** has some type of distortion because it must condense information into a small visual representation of something much larger.

DID YOU KNOW?

Maps allow for visual comparisons like population density between places and regions.

When reading a map, look at its title, the **scale** of measurement, and the **compass rose** indicating the cardinal directions of north, south, east, and west. Some maps show the intermediate directions of northeast, northwest, southeast, and southwest.

Look at the map's **key**, or legend, which indicates what the symbols, lines, and colors mean. Cities, capitals, and boundary lines may also be noted.

Small-scale maps show large areas like a country or the world. Large-scale maps show details of small geographical areas like cities.

Thematic maps show detail concerning a particular aspect of an area, like oil resources. Qualitative maps often show non-numerical information, such as locations of businesses, using colors, symbols, or lines. Quantitative maps illustrate numerical data such as numbers of crimes committed in a certain area. **Flow-line maps** show the movement of people, animals, goods, and ideas as well as physical processes like hurricanes or tsunamis. Arrows generally represent the flow and direction of movement.

Measuring Latitude

Lines of **latitude** are parallel lines of measurement that circle Earth horizontally. Parallels south of the equator are the South latitudes, and parallels north of the equator are the North latitudes. The North Pole is 90° N (North) latitude. The South Pole is 90° S (South) latitude.

The Equator

The equator is 0° latitude. Locations north of the equator lie in the Northern Hemisphere. Locations south of the equator lie in the Southern Hemisphere.

Measuring Longitude

Lines of **longitude**, or meridians, circle the earth vertically from pole to pole. They measure distance east or west of the **Prime Meridian**, which lies at 0° longitude, where the Royal Observatory stands in Greenwich, England. Meridians east of the Prime Meridian are called east longitudes, and meridians west of the prime meridian are the west longitudes.

TIP

The 180° longitudinal meridian on the opposite side of Earth from the Prime Meridian is the International Date Line.

Everything east of the Prime Meridian for 180° is in the Eastern Hemisphere. Everything west of the prime meridian for 180° is in the Western Hemisphere.

Absolute and Relative Location

Earth's maps are based on a grid system. Using the equator, Prime Meridian, and other lines of latitude and longitude on maps and globes helps us find the **absolute location** of places, or where the lines of latitude and longitude cross. Relative locations use well-known places as reference points. A relative location pinpoints the new place in relation to places already known, like landmarks, cities, rivers, lakes, states, or countries.

Time Zones

The creation of times zones across the world signified the beginning of the modern global era. Their establishment was tied to industrial development and the expansion of international trade.

Timekeeping was a local phenomenon before the late nineteenth century. Towns set their clocks according to the positions of the sun. Noon was defined as the time when the sun reached its maximum altitude above the horizon. Cities and towns would assign a clockmaker to calibrate a town clock to these solar positions. This town clock would then represent official time and the citizens would set their watches and clocks.

The second half of the nineteenth century was a time of increased human and technological movement. In the United States and Canada, large numbers of people moved west, and settlements in western areas expanded rapidly. To support these new settlements, railroads moved people and resources between the various cities and towns. However, because of the various ways in which local time was kept, the railroads experienced major problems in constructing timetables for the various stops. Timetables could only become efficient if the towns and cities adopted some type of standard method of keeping time.

In the 1870s, Sir Sandford Fleming of Canada suggested a system of worldwide time zones to simplify the keeping of time across the planet. Fleming proposed the globe be divided into 24 time zones, each 15 degrees of longitude in width—because the world rotates on its axis once every 24 hours, and there are 360 degrees of longitude, each hour of Earth rotation represents 15 degrees of longitude.

Railroad companies in Canada and the United States began using Fleming's time zones in 1883. In 1884, an international Prime Meridian conference was held in Washington, D.C., to adopt the standardized method of timekeeping and determine the location of the Prime Meridian. Conference members agreed the longitude of Greenwich, England, would become zero degrees longitude, or the Prime Meridian. The International Date Line (180 degrees longitude) marks the line where the date changes. It was also proposed that the measurement of time on Earth would be made relative to astronomical measurements at the Royal Observatory at Greenwich. This time standard was named Greenwich Mean Time.

Many nations today operate on variations of the time zones suggested by Fleming. In this system, time in the various zones is measured relative to the coordinated universal time (UTC) standard at the Prime Meridian. UTC is determined from primary atomic clocks coordinated by the International Bureau of Weights and Measures located in France.

National boundaries and political matters also influence the shape of time zone boundaries. For example, China uses a single time zone (eight hours ahead of coordinated universal time) instead of five different time zones.

DID YOU KNOW

Coordinated universal time became the standard legal reference for time all over the world in 1972.

Practice

Fill in the blank with the correct word from the information in the preceding paragraphs.

1. Geographers use ______________ such as aerial photography and satellite imaging to know more about mineral deposits or freshwater sources or to survey human activities on Earth.

2. Some maps show the __________ directions of northeast, northwest, southeast, and southwest.

3. A ___________ indicates what the symbols, lines, and colors on a map mean.

4. ___________ maps show detail concerning a particular aspect of an area, like oil resources.

5. Lines of ___________, or parallels, circle Earth in degrees horizontally along the equator.

6. The ___________ is 90° N (north) latitude.

7. The ___________ is 0° latitude.

8. The 180° meridian is the ___________.

9. ___________ location pinpoints the new place in relation to places already known.

10. Modern time zones were created relative to the ___________.

Interviewing

To find out what people think about a place, geographers may interview them. This will tell the geographer what people's beliefs and attitudes are about a place and why they believe what they do. Geographers select a carefully chosen representative sample from the area to determine what effects people's beliefs have on the physical environment, and vice versa.

Analyzing Statistics

Collecting and analyzing numerical information, like temperature and rainfall, over a long period of time helps identify a climate. Computers can help organize and present this information. Geographers look for patterns and trends in the numbers.

Using Technology

Modern **technology** allows cartographers to make maps with software programs. High-tech remote sensor cameras on orbiting satellites and radar can gather data and images on Earth's environment, weather, human settlement patterns, and vegetation.

New communications technologies like the Internet, cell phones, and overnight deliveries allow geographic information and goods to move faster than ever.

Geographic Information Systems

Geographic information systems (**GIS**) accept data from maps, satellite images, printed text, and statistics. The GIS converts the information into a digital code and places it in a database. Cartographers then program the GIS to process the data and produce maps.

DID YOU KNOW?

GIS can be used to track wildlife, plan roads, and create evacuation routes for natural disasters.

Through GIS, each type of information is saved as a separate electronic layer, allowing cartographers to make and change maps quite easily. The first layer might be a city. The second layer shows information on a specific problem like population density. A third layer might only deal with members of a certain interest group. A fourth layer might show a different or more specific group.

Careers in Geography

Geographic skills are useful in government, business, and education. Ecologists, for example, must know the geography of the area where they are studying wildlife. A travel agent has to know geography to plan trips for people.

Geography is useful for employment opportunities in teaching and education. You can teach geography in elementary, middle, and high schools and in colleges and universities. University research in geography has prepared many geographers for work in numerous industries.

There are many specialized fields of geographic work. Physical geographers can work as soil, weather, or climate experts. Geographers trained in environmental science are in demand as environmental managers and technicians. This work deals with assessing the environmental impact of proposed development projects on surrounding air and water quality and on wildlife. Physical geographers survey the land near the construction site and frequently prepare the environmental impact report often required before building can commence.

Jobs in healthcare, transportation, population studies, economic development, and international economics are available to trained geographers. Human geographers trained in urban planning often work for local and state government agencies. They handle housing and community development, parks and recreation planning, and urban and regional planning.

> **DID YOU KNOW?**
>
> Planners analyze land use and transportation systems and monitor urban land development.

Economic geographers examine human economic activities and their relationship to the environment. These geographers can work at market analysis and site selection for stores, factories, restaurants, or other businesses.

Regional geographers study the features of a particular region and could assist government and business to make decisions concerning land use. Geographers also find work as writers and editors for publishers of textbooks, maps, atlases, and news and travel magazines.

Geography and Other Disciplines

The tools and methods geographers use can help us understand historical, political, social, cultural, and economic aspects of problems dealing with the environment.

Historical Environments and Politics

Geographers may examine how human activities changed the natural vegetation of a place or how water, roads, or railways have changed the environment. Such inquiries may help us understand why things are as they are today and may help us plan for the future.

Geographers look at political boundaries and how and why they have changed. They look at how places are governed, and how the natural environment has influenced political decisions.

Society and Culture

Human or cultural geographers use sociology and anthropology to understand cultures. They examine the relationship between the physical environment and social structures and people's ways of life. They want to know how different group activities affect physical systems and how physical systems affect human systems differently.

Economies

Economies tell geographers how resource locations affect the way people make, use, and transport goods and how and where services are provided. Geographers want to know how locations are chosen for economic activities like mining, farming, manufacturing, and trade. Productive locations have plentiful resources and beneficial transportation routes.

Practice

Insert T or F to indicate whether the statement is true or false, based on what you've read.

11. ____ Representative sampling is important to geographers conducting interviews.

12. ____ Geographers can serve as environmental planners or managers.

13. ____ Cultural geographers deal with location and social structures.

14. ____ Geographers cannot specialize in urban planning.

15. ____ Geographers do not research how historical or political policies affect the environment.

16. ____ Geographical information systems can be used to track wildlife, plan roads, and create evacuation routes for natural disasters.

17. ____ Cartographers today use software mapping programs.

18. ____ High-tech remote sensor cameras on orbiting satellites and radar can gather data and images on Earth's environment, weather, human settlement patterns, and vegetation.

19. ____ Geographers work as writers and editors for publishers of textbooks, maps, atlases, and news and travel magazines.

20. ____ Cultural geographers avoid using sociology and anthropology to understand cultures.

Answers

1. Remote sensing
2. Intermediate
3. Key or legend
4. Thematic
5. Latitude
6. North Pole
7. Equator
8. International Date Line
9. Relative
10. Prime Meridian
11. T
12. T
13. T
14. F
15. F
16. T
17. T
18. T
19. T
20. F

LESSON

6 CANADA AND THE UNITED STATES—PHYSICAL GEOGRAPHY

LESSON SUMMARY

This lesson examines the physical geography—landforms, water systems, natural resources, climate, and vegetation—of the United States and Canada.

America forms the longest and straightest bone in the earth's skeleton.

—Huntington Ellsworth in *Red Man's Continent: A Chronicle of Aboriginal America* (1919)

Western Mountains, Plains, and Plateaus

The Rocky Mountains form the longest mountain range in North America, stretching for 3,000 miles from British Columbia, Canada, to New Mexico in the United States. The Rockies are young mountains created by tectonic activity. Some of its peaks reach over 14,000 feet.

The Pacific mountains include the Sierra Nevada and the Cascades, and the Coast, Olympic, and Alaska ranges. The highest point in North America is Mount McKinley in Alaska at 20,320 feet.

There are dry basins and plateaus between the Pacific mountain ranges and the Rockies. The Columbia Plateau covers parts of Washington, Idaho, and Oregon. The 6,000-foot deep Grand Canyon is part of the flat-topped mesas of the Colorado Plateau. The Great Basin, covering almost all of Nevada and parts of California, Oregon, Idaho, and Utah, includes Death Valley, the lowest place in the United States.

The Rockies slope down into the Great Plains. The Plains become the Central Lowlands along the Mississippi River.

Eastern Mountains and Lowlands

The Appalachian Mountains extend for 1,500 miles from the island of Newfoundland in Canada to central Alabama. They are North America's oldest mountains, eroded over time by running water, ice, and wind.

The Atlantic coastal plain runs east and south of the Appalachians, from New York to Florida. Between the mountains and the Atlantic coastal plain are the rolling hills of the Piedmont, stretching from New Jersey to Alabama. Rivers cut east through the Piedmont, making their way to the Atlantic Ocean. The Gulf Coastal Plain extends west to Texas and covers Mississippi, western Tennessee, Kentucky, and some of Alabama, Louisiana, and the Florida panhandle.

East of the Canadian plains, the mammoth Canadian Shield is a rocky core centered on the Hudson Bay and James Bay. It covers more than half of Canada and anchors the continent.

Islands

Newfoundland, Prince Edward Island, and Cape Breton Island in the east and Vancouver in the west are important to the Canadian economy. Only 13.4 miles long and 2.3 miles wide, New York City's Manhattan Island at the mouth of the Hudson River is a major U.S. and world economic center. The Caribbean islands, including the Bahamas, Cuba, and the Dominican Republic and Haiti, are part of North America, as well.

DID YOU KNOW?

The Hawaiian Islands are volcanic mountaintops.

Water Systems

Freshwater lakes and rivers helped make the United States and Canada prosperous. Abundant water transports resources and serves the needs of cities, rural areas, homes, and industries.

Rivers

The Continental Divide determines the flow of rivers. East of the divide, rivers flow east to the Atlantic Ocean, Arctic Ocean, and Hudson Bay. West of the Rockies, rivers flow to the Pacific Ocean. The Rockies are the source (or **headwaters**) of the Colorado and Rio Grande Rivers. These rivers connect many smaller rivers, or tributaries, and streams.

The Mackenzie River flows from the Great Slave Lake in Canada's Northwest Territories to the Arctic Ocean. It drains most of Canada's northern interior.

The Mississippi River flows 2,357 miles. Its headwaters are a tiny stream in Minnesota and its mighty mouth is 1.5 miles wide in Louisiana. It drains all or part of 31 states and two Canadian provinces into the Gulf of Mexico. Its enormous reach makes the Mississippi River one of the world's busiest commercial waterways.

Along the fall line where the higher land of the Piedmont drops to the lower Atlantic Coastal Plain, eastern rivers break into waterfalls and rapids, making ships unable to travel further inland. Towns along the fall line harness water power from waterfalls for mills and factories. In the northeast, cities along the fall line are ports for oceangoing trade.

Part of the border between the United States and Canada is formed by the St. Lawrence River. This river flows from Lake Ontario to the Gulf of St. Lawrence in the Atlantic Ocean. The Canadian cities of Quebec, Montreal, and Ottawa developed along the St. Lawrence River and its tributaries and depend on them for trade.

DID YOU KNOW?

Niagara Falls, on the Niagara River, also forms the border between Canada and the United States. The falls are a source of hydroelectric power for both countries.

Lakes and Other Waterways

Just like the Canadian Shield, the five Great Lakes were created by the gouging movement of glaciers. The Great Lakes basin links to the Atlantic Ocean via the St. Lawrence Seaway. A series of canals and channels, the Great Lakes-St. Lawrence Seaway System has been instrumental in America's economic development. Deposits of coal and iron in the area fueled industrial development. The seaway, which opened in 1959, made cities along the Great Lakes powerful trade and industrial centers. However, it also made the Erie Canal obsolete and led to the decline of many towns and cities in New York State.

Natural Resources

Abundant natural resources made the United States and Canada wealthy, but these resources and the areas where they are found need protection.

Fossil Fuels and Minerals

Coal, petroleum, and natural gas are fossil fuels. Fossil fuels are buried plant and animal remains that are hundreds of millions of years old. Because they are nonrenewable energy sources, fossil fuels must be conserved. Methods of extraction can damage the environment, so drilling for oil and mining for coal have become controversial issues.

There are petroleum and natural gas deposits in Texas, Alaska, and Alberta, Canada. The Trans-Alaska Pipeline, which was completed in 1977, has transported through 2010 some 16 billion barrels of **crude oil** across rough Alaskan terrain to the port of Valdez in southern Alaska. Coal is mined in the Appalachian Mountains, Wyoming, and British Columbia.

Mineral wealth is plentiful in North America. The Rockies have gold, silver, and copper. The Canadian Shield has iron and nickel and Michigan and Minnesota have iron. Canada also has a large portion of the world's gold, silver, and copper.

DID YOU KNOW?

Canada has 28% of the world's potash, or potassium carbonate, which is used to make fertilizer.

Minerals are nonrenewable and must be conserved. Because mining uses heavy equipment and large amounts of water, and moves rock and mineral, it can damage land, air, water systems, and ecosystems.

Timber and Fishing

Timber is plentiful in North America. Forest and woodlands cover the continent. However it is important to replant trees to replace those used for lumber, protect the thousands of species of native forest animals, and preserve old-growth forests.

The coastal waters of the Atlantic and Pacific oceans and the Gulf of Mexico are home to **fisheries**, and fishing remains important to regional economies. The Grand Banks off the coast of southeast Canada is one of the world's largest (109,000 square miles) and richest fishing grounds. Some coastal areas have been overfished, and Canada has banned cod fishing.

Practice

Fill in the blank with the correct word from the information in the preceding paragraphs.

1. New York City's ____________ Island sits at the mouth of the Hudson River.

2. The Grand Canyon is part of the flat-topped mesas of the __________ Plateau.

3. __________ is a source of hydroelectric power for the United States and Canada.

4. The Rockies slope down into the __________.

5. There are dry __________ and plateaus between the Pacific Ranges and Rockies.

6. The Rockies are the source, or __________, of the Colorado and Rio Grande Rivers.

7. The __________ are America's oldest mountains.

8. The Great Lakes basin links the Atlantic Ocean via the __________ Seaway.

9. __________, petroleum, and natural gas are nonrenewable fossil fuels.

10. Because __________ uses heavy equipment and large amounts of water, and moves rock and mineral, it can damage land, air, water systems, and ecosystems.

Climate and Vegetation

The climate of the United States is quite varied. The South has wet and dry seasons, high latitude areas can be bitter cold, the interior experiences extreme weather changes, and the Pacific coast tends to be cool and wet.

Southern Climates

Location near the coast and prevailing winds result in both warm, wet climates and warm, dry climates in the southern United States.

The humid subtropics of the Southeast have rainy, long, humid summers and mild winters, and water from the Atlantic Ocean prevents a dry season. In late summer and early autumn, the southeastern coast is pounded by **hurricanes**. The swamps and wetlands of the Florida Everglades teem with great varieties of vegetation and wildlife.

Hawaii in the Pacific and Puerto Rico in the Caribbean have tropical wet climates that support lush rain forests.

Hurricanes

Hurricane season lasts from June 1 to November 30. Hurricanes frequently form from high humidity, water temperatures over 80°F, and light winds, and are categorized from level 1 to level 5. Category 1 hurricanes have winds over 74 mph, and Category 5 hurricanes have winds over 155 mph. The eye of a hurricane is calm and it is surrounded by the strongest winds.

One of the largest hurricanes ever to hit the United States was Katrina in 2005. Katrina hit the Gulf coast states with 215 mph winds and 34-foot storm surges. It left thousands dead, missing, or homeless, and caused 80% of New Orleans to flood. Coastal towns were annihilated.

Warm and Dry Climates

When dry air moves down the leeward side of a mountain, the rain shadow effect creates a desert. The plateaus and basins between the Pacific Ranges and the Rocky Mountains stay hot and dry. Most of this area has a steppe or desert climate with blistering heat.

DID YOU KNOW?

Death Valley, at 134°F, had the highest temperature ever recorded in the United States.

Central and southern California have a Mediterranean climate with mild, wet winters and hot, dry summers. The vegetation there is drought-resistant woodland with twisted, hard-leafed trees. Mediterranean scrub, known as **chaparral**, is burned regularly to encourage growth, but widespread brush fires are a hazard around Oakland and Los Angeles when the hot, dry Santa Ana winds blow down mountain slopes from inland plateaus.

Northern Climates

Variations in climate and vegetation in the North result from the combined effects of latitude, elevation, ocean currents, and rainfall.

Interior Climates

The Great Plains in the center of North America has a humid continental climate with hot summers and bitterly cold winters. Western mountains block Pacific winds, but warm, moist winds blow north along the Rockies from the Gulf of Mexico. Cold, moist winds blow down from the Arctic. This climate reaches from the northeastern United States into southern Canada.

Naturally treeless grasslands called **prairies** spread across the Plains. Six-to-eight-foot-tall switch grass and blue stem grow an inch a day as they soak up 10 to 30 inches of rainfall a year. The Great Plains and northeast United States often receive violent spring and summer thunderstorms called **supercells**.

DID YOU KNOW?

These supercells sometimes spawn **tornadoes**, twisting funnels of air with winds up to 300 mph.

Coastal Climates

Ocean currents and westerly winds from the Pacific Ranges provide the Pacific coast from northern California to Southern Alaska with a Marine West Coast climate. The mountain barrier forces wet, warm air up and it cools to release moisture. Parts of the region thus receive more than 100 inches of rain per year. Winters are rainy and overcast. Summers are cool and cloudless. Ferns and mosses grow here.

High Latitude Climates

The Canadian Shield has a subarctic climate, and is home to coniferous forests. Parts of the United States and Canada located in the high latitudes experience a harsh, subarctic climate with high atmospheric pressure that causes frigidly cold winter winds. Parts of the United States experience winter blizzards with winds of more than 35 mph, heavy or blowing snow, and visibility of less than 1,320 feet.

From the subarctic Yukon Territory to Newfoundland, there are **mixed forests.** Lands along the Arctic coast are part of the tundra climate zone. Bitter winters and cool summers in this vast wilderness make the land inhospitable to most plants. Few people live here.

Practice

Insert T or F to indicate whether the statement is true or false, based on what you've read.

11. _____ A steppe climate is the same as a tropical wet climate.

12. _____ The northeastern United States receives Arctic winds.

13. _____ The eye of a hurricane is calm.

14. _____ The South has a wet and dry climate.

15. _____ The Pacific Coast has a cool and wet climate.

16. _____ Hurricanes can have winds from 74 mph to more than 155 mph.

17. _____ Wet air moving down the leeward side of mountains creates deserts.

18. _____ The Great Plains has warm winters.

19. _____ Tornadoes can have winds of up to 300 mph.

20. _____ The Yukon has a tundra climate.

Answers

1. Manhattan
2. Colorado
3. Ottawa
4. Great Plains
5. Basins
6. Headwaters
7. Appalachians
8. St. Lawrence
9. Coal
10. Mining
11. F
12. T
13. T
14. F
15. T
16. T
17. F
18. F
19. T
20. T

LESSON 7

CANADA AND THE UNITED STATES—HUMAN GEOGRAPHY

LESSON SUMMARY

This lesson deals with the human geography—population, culture, language, religion, economy, education, healthcare, arts, and family life—of the United States and Canada. The United States and Canada share the longest undefended border in the world. They share a democratic tradition, a similar way of life, and free trade.

Humankind has not woven the web of life. We are but one thread within it. Whatever we do to the web, we do to ourselves. All things are bound together. All things connect.

—Chief Seattle (1854)

U.S. Population Patterns

The United States is a country shaped by immigration. It has a continually shifting population and densely populated urban areas.

People

The American population is diverse—most of the 310 million people living in the United States are immigrants or the descendants of immigrants. Native Americans number about 2.5 million.

Many immigrants from Europe, Africa, Asia, and Latin America came to the United States seeking political and religious freedom and economic opportunity. Others were fleeing wars or natural disasters. Rich natural resources and rapid industrial development made America an attractive destination. With hard work, enthusiasm, and talent, immigrants overcame discrimination to spread their diverse cultures.

Density and Distribution

Outside of large urban settings, the American population is widely distributed. The great industrial and commercial centers of the Northeast and the Great Lakes are densely populated. People looking for a mild climate and economic opportunity also settle in clusters along the Pacific coast.

DID YOU KNOW?

The least populated areas of North America are subarctic Alaska, the parched Great Basin, and the arid Great Plains.

Since the 1970s, the mild climates of the American South and Southwest have grown the fastest. The **Sunbelt** states of California, Arizona, and New Mexico have developed their tourism, service, and manufacturing sectors.

Urban Areas

Cities, or urban areas, developed along waterways as mechanized farming displaced rural labor. Most Americans live in **metropolitan areas**—cities and their outlying suburbs—with populations that typically run into the millions. In the early to mid-twentieth century, as metropolitan areas became more crowded, **urban sprawl** fueled suburban development. This trend continues today.

The population of America's coastal port cities grew as they engaged in world trade. The Boston–Washington, D.C. **megalopolis**, or chain of large cities, supports huge populations. Over the years, Pacific coastal cities have been expanding their trade with Asian economies. Inland cities have grown up along rivers and lakes.

Early Nation Building

Although the early settlement of North America has been attributed solely to migration across an Asian land bridge some 20,000 years ago, recent evidence proves nomads may have arrived from South and Central America much earlier.

The Native Americans were shaped by their location and climate. Farmers of the desert southwest, for example, used irrigation to farm the dry land there.

As Europeans pushed west into the North American frontier from 1500 to 1900, Native American settlements were destroyed. The Spanish in the South established military posts, religious missions, cattle ranches, mines, and plantations using the natives as slaves. Frequent conflicts broke out in the backcountry between the Native Americans and the British settlers pushing into the interior.

The British controlled the Atlantic seaboard by the 1700s. Although New England had rocky soil and a short growing season, ample harbors and supplies of fish and timber made it a shipbuilding and fishing center. The fertile soils, mild winters, and warm summers of the Middle Colonies yielded **cash crops** for **export**. The mild climate and rich soil of the South yielded plantation agriculture.

The British government imposed taxes and restricted the freedom of colonists in North America. This led to the American Revolution (1775–1783) and the founding of an independent United States of America.

Growth, Division, and Unity

In the 1800s, the United States expanded far westward. Settlers overran Native American lands and traditional ways of life. Industry transformed the agrarian-based lifestyle. Factories harnessed water power in the Northeast to run machines in textile mills. Large supplies of coal from the Midwest fueled steam engines inexpensively. The United States became a leading industrial center.

The growing textile industry and the invention of the cotton gin made cotton the major Southern crop. Plantation owners cleared more land and the number of African slaves used for labor multiplied. Meanwhile, an informal network of secret safe houses called the Underground Railroad led some slaves to freedom in the North.

Tensions between the industrialized North and the agricultural South grew. After four years of bloody **civil** war (1861–1865), the industrial North emerged victorious. In 1865, slavery was outlawed by Amendment XIII.

Technological Change

The United States and Canada used vast energy and natural resources to industrialize.

The U.S. government encouraged settlement of the Great Plains to release pressure in crowded cities and to cultivate more food for the growing population. Farmers in the Plains caught and held rainwater through a technique called **dry farming**, which also involved planting drought-resistant crops like winter wheat. Steel plows and steam tractors made planting and harvesting large areas easier.

Immigrants from Ireland, Mexico, and China helped build railroad lines across the continent. Manufactured goods and food products could be shipped from east to west and north to south.

Two world wars made the United States a global power. Assembly lines cut costs and the time needed to make products. Standards of living increased and the population became more mobile and urban.

U.S. Culture

Immigration has brought cultural diversity to the United States.

Language and Religion

English is the main language of the United States. Some people are **bilingual** and speak a second language. Growing Latin immigration makes Spanish the second most widely used language in America.

Many religions are practiced in America. Christianity, Judaism, and Islam are the most prevalent religions.

Education and Healthcare

Young people must attend school in America until they are 16. The **literacy rate**, or number of people who can read and write, is 99%. Although America has an extensive public school system, many classrooms across the country are overcrowded. Many parents can no longer afford to send their children to college without incurring large debt.

Because of a highly developed economy, the United States can devote substantial resources to healthcare. Still, many people cannot afford any health insurance or healthcare.

DID YOU KNOW?

In 2010, a law was passed requiring all Americans to buy health insurance. The law further stated that people can no longer be denied health insurance due to preexisting conditions.

Arts

The Native Americans integrated art, music, and storytelling into daily life. With European settlement, arts were dominated by European traditions. By the mid-1800s, Americans began to create art forms to reflect their own lives and cultures. Writers started to write about life and culture in the different regions of the country. At the beginning of the twentieth century, **jazz,** which grew out of the work songs and spirituals sung by slaves, emerged.

American culture has permeated the world. Hollywood, a Los Angeles district, is home to the American movie business which has influenced culture around the world. Rock music has given the world musicians and forms as diverse as Elvis Presley (rock 'n' roll) and Bob Dylan (folk rock). Broadway in New York City is world renowned for first-class theater and musicals.

U.S. Family Life

Because America is one of the wealthiest countries in the world, its citizens have one of the highest standards of living in the world. Such a high socioeconomic status allows for a wide range of personal choices and opportunities. Food is relatively inexpensive because of an agricultural surplus.

Married couple families make up about half the households in the United States. Just under half of these families have children under 18 years old. Households today are smaller than in the past, with 60% having only one or two people. The average American is aged 36; in 1970, the average age was 28. The American population is living longer, which puts more strain on the social services sector.

Practice

Fill in the blank with the correct word from the information in the preceding paragraphs.

1. About half the immigrants in the United States today come from ____________.
2. The __________ rate tells you what percentage of people can read and write.
3. The ________ Americans integrated art, music, and storytelling into daily life.
4. ________ grew out of work songs and spirituals.
5. Because of a highly __________ economy, the United States can devote substantial resources to healthcare.
6. Some people are ______ and speak a second language.
7. Food is inexpensive in America due to an agricultural ____________.
8. Reflecting the richness of immigrant cultures, the American population is _________.
9. American households today are ___________ than in the past.
10. A wealthy country has a _________ standard of living.

Canada's Population Patterns

Canada is a highly developed nation with bustling cities and large, pristine wilderness areas. Canada has a diverse mixture of **ethnic groups**, densely populated urban areas, and vast expanses of rugged terrain. Some Canadians are descendants of Native Americans.

People

Immigrants came to Canada seeking political and religious freedom, economic and educational opportunity, and to escape war. **Loyalists**, colonists loyal to the British government, fled to Canada after the American Revolution. They settled the Maritime

provinces of Nova Scotia, New Brunswick, and Prince Edward Island.

TIP

A **province** is similar to a state.

Some immigrants settled in places where they could keep their way of life. French-speaking people, for example, settled in Quebec.

More than a third of Canadians identify themselves as having mixed ethnic origin; 1.3 million Canadians identify themselves as Native American, **Inuit** (Arctic natives of North America), or of mixed European and Native American ancestry.

Density and Distribution

Rugged terrain and a bitter cold climate make most of Canada inhospitable to human settlement. Most of the population lives along the U.S.-Canada border. The average population density is about 9 people per square mile, with more densely populated areas clustered around waterways and places that support agriculture.

Over the past 100 years, most migration in Canada has been west to the Prairie provinces of Manitoba, Saskatchewan, and Alberta. Oil and natural gas were discovered there in the 1960s.

Urban Areas

Most of Canada's population lives in urban areas. The industrial city of Toronto is Canada's largest city and its financial center. Montreal is an industrial and shipping center. Ships reach these inland cities using the St. Lawrence and Ottawa Rivers and the Great Lakes. Vancouver handles most of the trade between Canada and Asia. Edmonton grew with the petroleum industry.

History and Government

Canada's history has been influenced by Native roots, French and English culture, a drive for independence, and immigration.

Early Inhabitants and Settlers

The Vikings arrived on Newfoundland Island from Scandinavia about A.D. 1000. Englishman John Cabot arrived in Canada in 1497 in search of a shorter route to Asia. In 1534–1536, Frenchman Jacques Cartier explored the St. Lawrence as far south as Montreal, claiming it for France. Samuel de Champlain founded the first permanent European settlement in Canada at what is now Port Royal, Nova Scotia, in 1605. He also founded Quebec City in 1608.

TIP

The French organized their settlements in Canada and called them New France.

When Europeans first arrived in Canada in the 1400s, between 500,000 and 2,000,000 Natives were living there. As Europeans claimed land, diseases from Europe spread and the numerous native settlements quickly declined in North, South, and Central America.

In 1670, the British chartered Hudson's Bay Company to find the Northwest Passage to Asia, and operate a fur trade monopoly and settlements in Canada. Some Native tribes took sides as the British and French fought for control of trade in the area.

The British drove the French from the Hudson Bay area by capturing Quebec in 1759 and winning control of New France in 1763. The Quebec Act of 1774 gave French settlers the right to keep their language, religion, and legal system. The act also extended British-controlled Canadian territory south to the Ohio River. These land claims brought the

British into conflict with British-American land speculators.

United Canada

In the early 1800s, the British and French argued over colonial policy. They united against the fear of an American takeover. In 1867, the colonies of Quebec, Ontario, Nova Scotia, and New Brunswick united as provinces of the Dominion of Canada, a new country and part of the British Empire. Over the next hundred years, Manitoba, British Columbia, Alberta, Saskatchewan, Prince Edward Island, and Newfoundland joined the dominion as provinces.

Canada achieved full independence in 1931. The British government, however, retained the right to approve changes to the Canadian constitution. In 1982, Canada ended its legislative link to England. Canada is a constitutional **monarchy** today.

The executive branch of the Canadian government consists of the governor general, prime minister, and cabinet. The British monarch is still the head of state and appoints the governor general to serve in his or her place. The national legislature, **Parliament**, includes the Senate and House of Commons. Canada's prime minister is the head of government elected by the majority party in Parliament. Nine judges sit on Canada's highest court, the Supreme Court of Canada.

Expansion and Diversity

Throughout the 1800s, Canada annexed land from the Atlantic Ocean to the Pacific Ocean and from the Arctic Circle to the U.S. border. The British government encouraged emigration to Canada. From 1815 to 1845, one million people migrated from England to Canada, making the French population a minority for the first time. This sparked **nationalism** among the French-speaking Quebecois, a sentiment which would grow and resurface throughout Canadian history.

In the late 1800s, the fertile soils of the Prairie Provinces and the Klondike Gold Rush attracted immigrants from Germany, Scandinavia, the Ukraine, Japan, and China. Canada also sheltered escaped African slaves from the United States.

DID YOU KNOW?

Although slavery existed in Canada, it never reached the levels it did in the United States because the Canadian climate was unsuitable for plantation-style farming. Slavery was outlawed by British law in 1834.

Westward expansion in Canada pushed Natives off their lands. In 1998, the Canadian government formally apologized to Native peoples for mistreatment and established a healing fund for **reparations**. In 1999, the native Inuit won the right to their own territory, Nunavut, which means "Our Land."

Canada became a highly urban industrialized country in the 1900s. Minerals were extracted. Transportation systems and hydroelectric projects were developed. World War II stimulated the Canadian economy and made it a military and industrial power.

DID YOU KNOW?

Following the war, Canada expanded federal financial assistance to citizens through unemployment insurance, pensions, and medical care.

Modern Challenges

Trade is strong between Canada and the United States, but some Canadians do not like their cultural identity being threatened by American culture. French-speaking Canadians in Quebec and other provinces are pushing for increased protection of their language and culture. Many people in Quebec strongly support provincial independence or creation of a separate country apart from Canada.

Culture

European, Asian, and Native American cultures have influenced Canada's cultural diversity.

Language and Religion

French and English are the two official languages of Canada, reflecting the cultural struggle between settlers. British settlers brought the English language to most of Canada; in the province of Quebec, the French language prevails. German, Italian, Chinese, and the native Cree and Inuktitut or Inuit languages are also spoken.

Christianity is the most prevalent religion in Canada. The practices of Judaism, Islam, Buddhism, Hinduism, and Sikhism reflect the cultural diversity of Canada's immigrant population.

Education and Healthcare

Canada's network of public and private schools has produced a 99% literacy rate. Children must attend school from the ages of 6 to 16. Each province organizes and administers its own educational system.

Canada's public healthcare system is based on the principle that everyone deserves access to high-quality medical care. The federal government sets healthcare standards and each province finances and manages its own system. Because the Canadian life expectancy has increased, so have healthcare costs, so sometimes taxes have to be raised or benefits limited.

Arts

Native, British, French, and American cultures have influenced Canadian art. The various immigrant groups have added unique features to literature, visual arts, theater, and music. Since the 1950s, the Canadian government has encouraged locally produced cultural products for the national mass media.

Since 1900, scholars and museums have begun to appreciate the art of Native peoples. Influenced by European styles, Canadian painters have excelled in the painting of Canadian landscapes. The earliest Canadian literature was written by French explorers, missionaries, and settlers and had strong religious themes. Important twentieth-century Canadian writers include: Hugh MacLennan, a critic of contemporary Canadian life, and Margaret Laurence, who wrote about the prairies of central Canada. Margaret Atwood and Robertson Davies also have strong international followings.

Toronto is highly respected in music and theater. It is the third-largest film and television production center in the English-speaking world after New York City and London. The world-renowned Toronto Symphony Orchestra and top-ranked National Ballet of Canada are based in Toronto.

Family Life

Canada, with its high standard of living, is one of the wealthiest countries in the world. Almost everyone owns telephones or color televisions. About 56% of people own a car.

Canadian family and age structure are similar to the United States. Married couple families make up 69% of households in Canada and 60% of those have children in the home. Families are smaller than ever—43% have one child, while 39% have two children. Canadians are living longer. In 2001, one in six Canadians was over 65. In 1970, one in ten Canadians was over 65. Life expectancy in Canada is almost 81 years, which is among the highest in the world.

North America Today

Economy

The United States and Canada have **market economies**, where people own, operate, and profit from their own businesses. Market economies allow businesses to hire employees and pay them for their work. There are also laws to protect private property rights, employment opportunities, and the health and safety of workers.

Since 2007, the United States housing market and related industries crashed, triggering a huge recession that many observers call the worst economic catastrophe since the global depression of the 1930s. Toward the end of 2010 almost 10% of Americans were unemployed, and more than a quarter of Americans were underemployed. Food stamps recipient numbers skyrocketed in the states in 2010.

Postindustrial Economies

Most of the economic growth in the United States and Canada is occurring in **service industries**. About 75% of workers in the United States and Canada are employed in service industries such as healthcare, education, government, and banking. Much of the **postindustrial** economy is also heavily based on high-tech industries. While agriculture and manufacturing depend on access to natural resources and transportation, high-tech industries depend much less on transportation.

Manufacturing

Manufacturing makes up about 13% of the Canadian and 12% of the American economies. Robotics and computerized automation have transformed manufacturing in the region. Factories now produce more goods with fewer workers.

The late 1990s trend away from **heavy industry** left cities east of the Great Lakes without their economic base. Companies moved south to the Sunbelt. Many industrial areas, factories, and mills were abandoned and the area became known as the Rust Belt. However, many cities in the area have **retooled** old factories for use in new industries.

Most of the manufacturing and exports from the region consist of machinery and transportation equipment. California and Washington produce aircraft and aerospace equipment. The Midwest assembles automobiles. California and the Northeast process food. Canada, particularly Quebec, manufactures and exports large quantities of wood-based products, using their timber resources.

Agriculture

Most farming in the United States and Canada is **commercial**, with agricultural **commodities** produced for sale. Most farms are owned by families with **cooperative** operations.

The United States uses about 900 million acres of land for agriculture. Canada has much less arable land suitable for farming, but uses about 167 million acres for farming.

Although the number of farmers in the United States and Canada has decreased, the size of farms has increased since the 1950s. The decline is the result of the high cost of farming, the time and hard work required to run a farm, and natural disasters.

Key Agricultural Products

- Cattle ranches span the western, southern, and midwestern United States and the Prairie provinces of Canada. The north central United States, Ontario, and Quebec raise livestock.
- Wheat is grown on the Great Plains, or Wheat Belt, of the United States and in the Prairie provinces of Canada.
- The Corn Belt stretches from Ohio to Nebraska. Corn is also grown in Ontario, Manitoba, and Quebec.

Agricultural Technology

Geographic factors have limited where certain types of agriculture can be based. Cattle ranching needs wide open spaces and the natural grasses of the western plains and prairies.

Most American dairy farms are situated from upstate New York to Minnesota. Known as America's Dairyland, it has cooler summers and native grasses ideal for dairy cattle.

When breeds of cattle were developed that needed less land to graze, the southern United States opened up to cattle ranching.

NAFTA

In 1989, the United States and Canada signed the United States–Canada Free Trade Agreement (FTA) which removed trade restrictions between them. The 1994 **North American Free Trade Agreement** (NAFTA) includes the United States, Canada, and Mexico. NAFTA created one of the world's largest trading blocs. It eliminated trade barriers, increased economic activity among the three countries, and strengthened their positions in the global economy.

Developed countries like the United States have recently changed the geography of production and manufacturing. They have sought lower production and labor costs by **outsourcing**, or setting up operations with plants, service industries, and other businesses abroad to assemble products for domestic use or sale.

DID YOU KNOW?

NAFTA does not allow the free flow of labor from one country to another; it does allow U.S. companies to set up assembly plants in Mexico where labor costs are lower.

Managing Resources

People in the United States and Canada are evaluating the negative effects of human activity on the environment. They are also realizing the importance of wisely managing natural resources.

Taking out whole forests, known as **clear-cutting**, has destroyed many of the old-growth forests, left the land susceptible to erosion and flooding, and endangered wildlife.

Many wetland areas like swamps, marshes, and ponds have been disappearing due to conversion to urban or agricultural land use and pollution. Wetlands protect important water supplies and fisheries and often buffer coastal areas from storms and flooding.

DID YOU KNOW?

Levees, or raised embankments, built around New Orleans have destroyed nearby wetlands that once protected the area from flooding.

Ignoring the balance between species in ecosystems is another form of resource mismanagement. Many fisheries have been depleted. **Overfishing** causes the amount caught to exceed the amount resupplied by natural reproduction. The hunting and driving away of certain wildlife, like wolves, in western Canada has led to an overpopulation of other types of wildlife, like elk.

The deliberate or accidental introduction of nonnative plant or animal species due to global trade and travel causes many environmental problems like blocked waterways, displacement of crucial native species, and crop destruction.

DID YOU KNOW?

Efforts have begun to reverse the environmental damage, but there is a long way to go to maintain a sustainable level of natural resources.

Smog

When the sun's rays interact with automobile exhaust gases and industrial emissions, a visible haze called smog can form. The same chemicals and water vapor that create acid rain form smog. Smog damages or kills plants and harms people's eyes, throats, and lungs.

Some major cities have substantially reduced air pollution using clean air practices. Car manufacturers are producing fuel-efficient hybrid vehicles that have an electric and a gas-powered motor that work in concert.

DID YOU KNOW?

Research is being done on fuel cell vehicles that produce electricity using hydrogen fuel and oxygen, on biofuel vehicles that use fuel from organic sources like plant oils, and on solar-powered cars.

Water Pollution

Sewage and industrial and agricultural wastes have polluted North America's water systems. Industrial wastes are illegally dumped into rivers and streams or leak into the groundwater. The release of industrial wastewater into cooler lakes and rivers results in thermal pollution. Runoff from chemical-based agricultural fertilizers and pesticides also pollutes waters.

Water pollution speeds up **eutrophication**, the dissolution of trace nutrients, encouraging excess algae growth. Algal blooms deplete oxygen in water needed by other organisms once the algae die and decay.

DID YOU KNOW?

Algae can turn a lake into a marsh and later dry land.

Reversing the Effects of Pollution

In 1969, the U.S. Congress passed the National Environmental Policy Act (NEPA), which mandates that the government monitor the effects of its policies on the environment. The U.S. Environmental Protection Agency (EPA) was founded in 1970 as a research and watchdog agency to safeguard America's air, land, and waters.

DID YOU KNOW?

The amount of carbon dioxide in the atmosphere is the highest it has been in thousands of years.

Global Warming

The Arctic regions of Canada and Alaska demonstrate the effects of global warming. The caribou, polar bears, and seals have had to move north because of thinning ice. The Inuit who depend on the hunting of these animals have had to move north, too, to preserve their traditional way of life and simply survive. The permafrost is starting to thaw, bend the land, and weaken house foundations. Whole villages have sunk.

To combat global warming and greenhouse gases, the United States and Canada are attempting to lessen their dependence on fossil fuels. Both governments are offering subsidies to utility companies for the use of renewable energy sources. Solar panels and biofuels made from corn and other organic sources are potential energy sources. However, some people wonder whether the dependability, efficiency, and cost of other energy sources make them worthwhile.

Practice

Insert T or F to indicate whether the statement is true or false, based on what you've read.

11. _____ There are not any viable alternative energy sources besides fossil fuels.

12. _____ There is not any solid physical evidence to prove that global warming exists.

13. _____ Canada and the United States are considered developed countries.

14. _____ Fertilizers and pesticides are no threat to the water supply.

15. _____ Smog not only harms people's eyes, throats, and lungs, but also kills plants.

16. _____ Ignoring the balance of species in an ecosystem is an example of resource mismanagement.

17. _____ Clear-cutting whole forests is a good idea.

18. _____ The United States and Canada have extensive prairie lands and farmlands.

Answers

1. Latin American
2. Literacy
3. Native American
4. Jazz
5. Developed
6. Bilingual
7. Surplus
8. Diverse
9. Smaller
10. High
11. F
12. F
13. T
14. F
15. T
16. T
17. F
18. T

LESSON

8

LESSON SUMMARY

This lesson deals with the landforms, water systems, natural resources, climate, and vegetation of Latin America. Latin America includes South America, Central America, the Caribbean islands, and Mexico.

If man doesn't learn to treat the oceans and the rain forest with respect, man will become extinct.

—Peter Benchley

Countries

Antigua and Barbuda
Argentina
Aruba
Bahamas
Barbados
Belize
Bolivia
Brazil
Cayman Islands
Chile
Colombia
Costa Rica
Cuba
Dominica
Dominican Republic
Ecuador
El Salvador
French Guiana
Grenada
Guadalupe
Guatemala
Guyana
Haiti
Honduras
Jamaica
Martinique
Mexico
Nicaragua
Panama
Paraguay
Peru
Puerto Rico (U.S.)
St. Barthélemy
St. Kitts and Nevis
St. Lucia
St. Vincent and the Grenadines
Suriname
Trinidad and Tobago
Turks and Caicos
Uruguay
Venezuela
Virgin Islands (U.S.)

Landforms

South of the United States in the Western Hemisphere, Latin America takes up about 14% of Earth's land surface and covers approximately 8 million square miles. Latin America has four subregions: Mexico, Central America (from Belize and Guatemala to Panama), the Caribbean, and South America.

Mountains and Plateaus

The Andes mountains are part of the American **cordillera**, a belt of continuous mountain ranges that stretches along the western spine of North and South America. Running 4,300 miles from Venezuela and Colombia to southern Chile, it is the world's longest mountain chain, and it contains some of the highest peaks outside of Asia, including Mt. Aconcagua in Argentina, which exceeds 22,800 feet. The American cordillera also includes the Sierra Madres in Mexico and the Central Highlands in Central America.

Latin America is located on the Pacific Ring of Fire, a 25,000-mile horseshoe-shaped zone where for millions of years plates in Earth's crust have been colliding. These collisions formed mountains and volcanoes and still cause earthquakes.

Humans have settled in the mountains and plateaus for thousands of years. The cool mountain climates and rich natural resources like water, volcanic soil, minerals, and timber attracted settlers.

The rugged terrain isolated communities but new media such as television, cell phones, and the Internet are breaking down physical barriers.

Mountains of Central America and the Caribbean

The densely populated Mexican plateau is surrounded by the Sierra Madre mountains. Mild climate, volcanic soil, and adequate rainfall attracted settlers there for thousands of years.

Farther south, the volcanic peaks of the Central Highlands cross Central America and extend into the Caribbean Sea all the way to the islands.

The Caribbean islands are volcanic mountains that are part of the Central Highlands. Some volcanic peaks on Caribbean islands are still active. The Caribbean islands also often face powerful hurricanes that form in the Atlantic Ocean.

The Andes of South America

Tectonic activities created the Andes. Plate movements still cause earthquakes and volcanic eruptions.

The rugged terrain of the high Andes makes movement difficult. Cordilleras keep settlements isolated from each other. Many of these villages maintain centuries-old Native traditions.

In Peru and Bolivia, high plains, or **altiplano**, are encircled by the Andes, spanning huge areas between the two major mountainous areas. In southern Argentina, hills and flatlands form the plateau of Patagonia. The Andes to the west keep Patagonia dry, barren, and windy. Such wide plateaus were carved out by intense volcanic activity.

Highlands of Brazil

Eastern South America has broad plateaus and valleys. The Matto Grosso Plateau in Brazil, Bolivia, and Peru is a sparsely populated forest and grassland area. Farther east, the Brazilian Highlands spans several climate and vegetation zones. With wide open spaces and warm climate, the Brazilian Highlands is used to raise livestock. The eastern highlands **escarpment** steeply slopes down, plunging into the Atlantic Ocean.

Lowlands and Plains

Narrow coastal lowlands lie along the Gulf of Mexico, the Caribbean, and the Atlantic and Pacific Oceans.

Brazil has a long coastal plain. It starts in the country's northeast region and extends south to Uruguay. Inland development is difficult, with the escarpment rising from the coast into the highlands. Most of Brazil's population, therefore, lives along the coast.

The inland grasslands, or **llanos**, of Columbia and Venezuela and the **pampas** of Argentina, Uruguay, and southern Brazil are cattle grazing grounds. Ranchers employ cowboys known as gauchos to drive herds across the rolling plains. The fertile soil of the pampas produces wheat and corn and is one of the world's breadbaskets.

Water Systems

Latin America has an expansive river system. The Amazon is the longest river in the Western Hemisphere and second longest in the world, next to the Nile. It flows 4,000 miles from its headwaters in the Peruvian Andes through the heart of South America. The Amazon is an important transportation route from the Atlantic into Brazil's interior. Ships travel 2,300 miles upstream from the Atlantic Ocean in the navigable parts of the river.

DID YOU KNOW?

The Amazon is more than ten times more powerful than the Mississippi River in terms of water volume per second.

Hundreds of smaller rivers join the Amazon to form the Amazon basin, which drains an area of 2 million square miles. In Brazil, the Amazon basin drains into the Atlantic Ocean.

The Paraná, Paraguay, and Uruguay rivers form the second-largest river system in Latin America. This system provides important commercial transportation routes and is a major source of hydroelectric power.

After coursing through inland areas, the three rivers flow into a broad **estuary** where the ocean tide meets a river current. This estuary, the Rio de la Plata ("River of Silver") flows into the Atlantic Ocean, where it drains the rainy eastern half of South America.

In contrast to the giant rivers of South America, Central America's rivers are usually small. The Rio Grande, known in Mexico as the Rio Bravo del Norte, forms part of the border between Mexico and the United States.

The man-made Panama Canal is an important waterway. Built across the Isthmus of Panama, the canal provides a much shorter route between the Atlantic and Pacific oceans than the much longer route around South America's Cape Horn.

Latin America does not have many large lakes. Lake Titicaca in the Andes of Bolivia and Peru is the world's highest navigable lake, at 12,500 feet above sea level. Although actually an inlet of the Caribbean Sea, Lake Maracaibo in Venezuela is South America's largest lake. The lake itself and the area around it contain Venezuela's oil fields.

Natural Resources

Latin America has abundant natural resources but faces obstacles to development.

There are large oil and gas deposits along the Gulf of Mexico and in the southern Caribbean Sea. Few countries besides Mexico and Venezuela benefit economically from these energy resources. Venezuela has 67% of the region's oil reserves and Mexico has 13%.

The region is also rich in mineral resources. Venezuela has large amounts of gold, and Peru and Mexico have silver. Mines in Colombia have produced the finest emeralds for a thousand years. Chile is one of the world's largest copper exporters and Jamaica is a leading producer of bauxite, the main source of aluminum.

Because of diverse landforms, the region's resources are unevenly distributed. Inaccessibility, lack of development capital, and deep social and political divisions keep many resources from full development. The regional challenge is how to overcome such obstacles and best use the natural resources.

Practice

Fill in the blank with the correct word from the information in the preceding paragraphs.

1. Latin America has four subregions: Mexico, ____________ America from Belize and Guatemala to Panama, the Caribbean, and South America.

2. The ____________, located on the west side of South America, is the world's longest mountain chain.

3. ____________ and Venezuela have most of Latin America's oil reserves.

4. The Rio ____________, or Rio Bravo del Norte, in Mexico forms part of the border between Mexico and the United States.

5. Lake ____________ in the Andes of Bolivia and Peru is the world's highest navigable lake, at 12,500 feet above sea level.

6. An estuary, the Rio ____________, the "River of Silver," flows into the Atlantic Ocean where it drains the rainy eastern half of South America.

7. The man-made ____________ Canal cuts the distance for trade between the Atlantic and Pacific oceans.

8. The inland grasslands, or ____________, of Columbia and Venezuela are used for cattle grazing.

9. Most of Brazil lives along the ____________.

10. Lake ____________ contains Venezuela's oil fields.

Climate and Vegetation

Latin America has diverse climates, from steamy rain forests with millions of biodiverse species of plants and animals living together, to grassy plains, arid deserts, and sandy beaches.

DID YOU KNOW?

Location and landforms create vertical climate zones and tropical areas in Latin America.

Tropical Wet

Most of Latin America has a tropical wet climate with lush tropical rain forest vegetation. The region is located near the equator, so temperatures are high. Warm moist air is carried by the prevailing winds of the Atlantic Ocean, so there is abundant rainfall all year.

The Amazon rain forest covers one-third of South America. It is the world's largest tropical rain forest. Trees grow close together and form a dense canopy as high as 150 feet, blocking sunlight from reaching the forest floor. The Amazon absorbs heavy rains throughout the year.

DID YOU KNOW?

The Amazon shelters more species of animals and plants per square mile than any place on Earth.

The Amazon basin is the world's wettest tropical plain. Heavy rains drench most of the densely forested lowlands all year. During the rainy season, the sediment-rich Amazon River frequently floods.

Human activities have modified forest cover and land use patterns. The Amazon rain forest is facing heavy **deforestation** and conversion into large commercial plantations and cattle ranches.

Tropical Dry

Most Caribbean islands, north-central South America, and the coast of southwestern Mexico have a tropical dry climate. These areas have high temperatures, abundant rainfall, and an extended dry season. Grasslands flourish in many tropical dry areas. Some of these grasslands, like the llanos of Colombia and Venezuela, have scattered trees. They are often transition zones between grasslands and forests. Tropical dry soils are not very fertile or useful for large-scale agriculture.

DID YOU KNOW?

Water supply and flood control projects have made the llanos into fertile farmland.

Humid Subtropical

Most of southeastern South America has a humid subtropical climate. Winters are short and temperatures are cool to mild. Summers are hot, humid, and long. Rainfall is light year round but can be heavier in summer.

The natural vegetation of the humid subtropical climate is short grasses. The large forest groves were cleared for cattle ranching by Spanish settlers and overgrazing left only short clumps of grass to anchor the soil. To hold the topsoil, farmers plant alfalfa, corn, and cotton crops.

Beef cattle still graze on the plains and grasslands of the pampas. The rich soil of the pampas yields wheat and corn for global export.

Dry Climates

The southeastern coast of Argentina, coastal Peru and Chile, and parts of northern Mexico have desert climates and vegetation. Shifting winds and the cold Peru Current (also known as the Humboldt Current), a northwestern-flowing current along the western coast of South America, create dry coastal deserts.

DID YOU KNOW?

The coastal Atacama Desert is so arid that in some places rainfall has never been recorded.

Areas of vegetation called lomas thrive in the desert because of fog near the coast. In other Latin American deserts, soil is poor and vegetation sparse. Drought-resistant shrubs and prickly cacti have adapted to the specific conditions of this harsh environment.

Northern Mexico, northern Brazil, and south-central South America receive little rainfall. They do not have desert climates and vegetation. They have steppe climates, with hot summers, cool winters, and light rainfall. The natural vegetation is grassy or lightly forested.

Elevation and Climate

Although located in the tropics, some areas of Latin America are more affected by elevation than by distance from the equator. As altitude increases, soil, crops, livestock, and climate change.

Spanish terms describe five vertical climate zones found in the highlands of Central America and western South America. Differences in elevation define each climate zone.

The **tierra helada**, or "frozen land," lies above the snow line at around 15,000 feet. In the tierra helada, the temperature is less than 20°F. Snow and ice remain permanently frozen on the peaks of the Andes.

The **puna** climate zone, above the tree line at 12,000 feet, is very cold. The temperature ranges from 20° to 55°F. The puna has some grasses suitable for grazing sheep, llamas, and alpaca. There are no trees here.

From 6,000 to 12,000 feet is the **tierra fria**, or "cold land." Here temperatures range from 55° to 65°F. Winter frosts, widely spaced evergreen trees, and dense scrub are common in the tierra fria. Potatoes and barley grow well here. These cooler climates and natural resources attract human settlement. Some of Latin America's largest cities like Bogotá, Colombia, and Mexico City, Mexico, are located in this zone.

Between 2,500 and 6,000 feet lies the **tierra templada** or "temperate land." Here temperatures range from 65° to 75°F. At low altitudes, there are broad-leafed evergreens. At higher elevations, there are needle-leafed cone-bearing evergreens. This is the most densely populated climate zone. Coffee and corn are the main crops.

The **tierra caliente**, or "hot land," extends from sea level to 2,500 feet in elevation. Bananas, sugar, rice, and cacao are the main crops of the tierra caliente rain forests.

Practice

Insert T or F to indicate whether the statement is true or false, based on what you've read.

11. ____ The tierra caliente has some grasses suitable for grazing sheep, llamas, and alpaca.

12. ____ Bananas, sugar, rice, and cacao are the main crop of the tierra caliente rain forests.

13. ____ The tierra helada, or "frozen land," lies at 6,000 to 12,000 feet and has rain forests.

14. ____ The Amazon basin is the world's wettest tropical plain.

15. ____ The vegetation in the pampas consists of short grasses.

16. ____ Most of Latin America has a tropical wet climate with tropical rain forest vegetation.

17. ____ The Atacama Desert receives seasonal rains.

18. ____ The Amazon shelters more species of animals and plants per square mile than any place on Earth.

19. ____ The tierra templada is the most densely populated climate zone in Latin America.

20. ____ The southeastern coast of Argentina, coastal Peru and Chile, and parts of northern Mexico have desert climates and vegetation.

Answers

1. Central
2. Andes
3. Mexico
4. Grande
5. Titicaca
6. de la Plata
7. Panama
8. Llanos
9. Coast
10. Maracaibo
11. F
12. T
13. F
14. T
15. T
16. T
17. F
18. T
19. T
20. T

LATIN AMERICA—HUMAN GEOGRAPHY (PART I)

LESSON SUMMARY

This lesson examines the human geography, or population, culture, language, religion, economy, education, healthcare, arts, and family life, of Mexico, Central America, and the Caribbean.

The Gateway of the Sun in Tiahuanaco [Bolivia] may date back to the eleventh millennium B.C., the same time the Sphinx was originally carved. This means Tiahuanaco may have been influenced by a lost civilization, Atlantis more or less. We must make sense of the astonishing cultural similarities between Mexico, South America, and Egypt.

—Oswaldo Rivera, Director, Bolivian National Institute of Archaeology, in Graham Hancock's *Quest for the Lost Civilization* (1998)

Mexico

Mexico has been shaped by the Maya and Aztec civilizations, as well as by Spain.

Population Patterns

Ethnic groups, migration, and urban growth have shaped population in Mexico. Most of the people who settled here in ancient times, such as the Aztecs, settled in the Mexican plateau in northern and central Mexico.

People

Starting in the Age of Exploration in the early sixteenth century, most Natives in Mexico died from European diseases, not war. The Natives who remained mixed with the Europeans and had biracial, or **mestizo**, children.

Density and Distribution

Although Mexico is the most populous Spanish-speaking country, with approximately 111 million people, its population density is 147 people per square mile—not very crowded. However, Mexico City, with roughly 21 million people in 573 square miles, is extremely crowded, averaging 36,650 people per square mile.

Many Mexicans have left rural areas for the city because of limited land and their desire for social services and economic opportunities. Many Mexicans move near the U.S. border seeking work. Seventy-five percent of Mexicans now live in cities.

Urban Areas

Growing cities often absorb surrounding cities and suburbs to create **megacities** with populations of more than 10 million people. Mexico City is a **primate city**, a country's leading city that dominates its economy, culture, and political affairs, and usually has a population at least twice the size of the next largest city.

History and Government

Ancient Mexico

The indigenous people of Latin America say their ancestors came across the seas following a great flood. Their oral history contradicts the long-standing theory that early migrations to Latin America originated with Asians 14,000 to 20,000 years ago, crossing a land bridge from Asia. Recently discovered artifacts demonstrate another hypothesis, that ancient people may have migrated across the Pacific Oceans 20,000 to 30,000 years ago.

Giant heads in African or Polynesian styles left by the Olmec or their ancestors in La Venta, Mexico date back to at least 3500 B.C. These objects have led many scholars over the past two decades to question the Asia-only theory of the settlement of Mexico.

The Maya, who settled the Yucataán Peninsula of Mexico, created an agricultural and trading empire from A.D. 250 to 900. The Catholic Spaniards believed the Maya were devil worshippers because they practiced human sacrifice on temple grounds.

The Maya made **glyphs**, or symbols often written on stone, to honor their gods and record history on temple walls. As brilliant at mathematics as the Babylonians (who had a strikingly similar numerical system), the Maya invented one of the most elaborate universal clocks ever created. No one knows what happened to the Maya; they appear to have just abandoned their cities.

At least 40 Mayan cities have been discovered, but most of their writing has not been translated. Descendants of the Maya remain in villages in southern Mexico and Central America, where they practice subsistence farming.

Another indigenous people, the Aztecs, reached the height of their power in the late 1400s. Their name came from Aztlán, the mythical place they believed their ancestors came from. They built their capital on the island of Tenochtitlan, which today is buried under overpopulated Mexico City.

The ancient peoples of Mexico, Central America, and South America seem to have had an in-depth understanding of the orbits of the sun, moon, planets, and stars. They constructed their temples to reflect this highly advanced scientific knowledge of the universe.

DID YOU KNOW?

The Aztec grew crops on **chinampas**, floating islands made from large rafts covered with mud from the lake bottom.

When the Spanish **conquistador** (conqueror) Hernán Cortés arrived in the Aztec capital in 1519, he noted it was an imperial city of over half a million people with bustling canals used for trading with client states. The Aztec capital also had opulent pyramid temples where human sacrifices, including the removal of beating hearts, were practiced. Cortés commented that the Aztec capital was grander than any city in Spain at the time.

The Aztecs had a highly structured class system headed by an emperor and military officials. High-ranking priests performed blood rituals to win the favor of the gods. Most Aztec people were farmers, laborers, and soldiers.

Independent Nation

Cortés slaughtered the last of the Aztecs in 1521 and claimed Mexico for Spain. Mexico remained part of the viceroyalty of New Spain for 300 years.

In the 1700s, resentment against European rule escalated. The American and French revolutions were particularly influential. The first Spanish-ruled country in Latin America to declare and win independence was Mexico. Father Miguel Hidalgo led Mexico's struggle for independence, which lasted from 1810 to 1821.

A small, elite group of wealthy landowners, army officials, and clergy seized economic and political power in Mexico. In this time of power struggles, public disappointments, and chaotic, outlaw-styled revolts, the **caudillo**, or absolute dictator, came to power with the backing of the wealthy landowners and military.

A new constitution established in 1917 brought reforms to the people and made Mexico a **federal** republic.

Starting in 1929, the Partido Revolucionario Institucional (PRI) dominated the presidency and Mexican politics for nearly 70 years. The PRI's control ended in 2000, when Vicente Fox of the opposition party Partido Acción Nacional (PAN) won the presidency. Felipe Calderón won the presidency for the PAN again in 2006.

Today, struggles for reform and political power in Mexico continue. Native Americans, workers, and farmers continue to pressure the government for greater inclusion in the political system.

Language and Religion

The official language of Mexico is Spanish. Ninety-five percent of the population speaks Spanish, though another 62 indigenous languages are still spoken there. Nearly 83% of the population is Roman Catholic, though many indigenous people retain aspects of their traditional beliefs. Mixing religious beliefs into one faith is called **syncretism**.

Education and Healthcare

Education varies significantly across Mexico. Most public schools are in rural areas, but they lack the funding and qualified teachers of urban or private schools. Some government promotion of adult literacy and school funding has helped.

If employment and education improve, health problems linked to poverty, lack of sanitation, and malnutrition will decrease. Because the federal government subsidizes healthcare, it is available to all citizens. However, the poor quality of public medicine in rural areas leads many Mexicans to seek medical care in cities or in other countries.

Trade

The 1994 **North American Free Trade Agreement (NAFTA)** reduced trade restrictions and increased the flow of goods, services, and people. Trade among the United States, Mexico, and Canada has increased by 10 to 15%, but effects on employment and gross domestic product (GDP) have been small. American labor has opposed NAFTA because Mexican workers will accept lower pay. Few American companies have moved to Mexico, due to high production and electricity costs. Mexico has increased its exports and received more international investment and jobs, but NAFTA has not helped the poor people of Mexico much.

Arts

Mexico's art shows indigenous and Spanish colonial influences. Early Native American architecture includes Maya pyramids and Aztec temples. Some of these early Native masterpieces of architectural and cosmic genius were decorated with murals or wall paintings and mosaics. Churches and other buildings reflect European architectural styles.

The twentieth century brought a revival of interest in precolonial history and culture. For example, Diego Rivera's **murals** and frescoes illustrate Mexico's history and culture in vivid detail. Other outstanding Mexican painters include Frida Kahlo and Clemente Orozco. Mexico's past has inspired the writers Octavio Paz and Carlos Fuentes. Ballet Folklórico performs Native American and Spanish dances.

Family Life

Family life is important to Mexicans. Parents and children often share their home with members of an extended family. Like most Latin American societies, Mexican society still displays elements of machismo, a Spanish and Portuguese tradition of male supremacy. However, women have made rapid advances in recent decades.

DID YOU KNOW?

Compadres, or godparents, are chosen by parents to sponsor their new baby and watch over his or her upbringing.

Sports and Leisure

Spectators crowd arenas to watch bullfighting, Mexico's national sport. People are also passionate about fútbol (soccer). Baseball and jai alai, which is like a cross between handball and Native-born lacrosse, are also very popular. People in Mexico love to celebrate. From friendly gatherings to special family dinners, religious holidays, and patriotic events, almost any social occasion is a party, or *fiesta*.

Central America and the Caribbean

Population Patterns

Diverse ethnic groups, migrations, small land areas, and rapid growth have shaped the population of Central America and the Caribbean.

Indigenous cultures are often mixed with those of Spanish settlers in Central America and the Caribbean. Mixed in, too, are English, French, African, Dutch, East Indian, Chinese, and other cultures.

People

The first inhabitants of Central America and the Caribbean were indigenous peoples. People of Maya descent make up about half the population of Guatemala. Most of the people in Costa Rica are descendants of Europeans. At least two-thirds of the people in Central America are mestizo.

Europeans brought Africans here as slaves by force. Slavery ended in the region in the 1800s, but many Africans who lived in Latin America for generations stayed. In the Bahamas, most of the people are of African descent. Other parts of the Caribbean, like the Dominican Republic and Cuba, blend African and European ethnicities.

DID YOU KNOW?

People of mixed African and European ancestry are called mulattos.

Density and Distribution

Most people in Central America live in the highlands that run mostly along the Pacific Coast. Population densities vary, however. Guatemala's population density of 342 people per square mile is about ten times that of Belize, which has 36 people per square mile.

Population density is heavy in the Caribbean, which has small land areas with large populations that grow at rapid rates. There are fewer people per square mile in Central America.

Since the 1970s, large numbers of people have been leaving Central America and the Caribbean for better external economic opportunities and to escape civil war or unstable political situations.

Internally, the people of Central America and the Caribbean have been moving to cities.

Urban Challenges

City resources are strained by rapid population growth. Jobs and housing may be scarce. City **infrastructure** can collapse, depriving people of electricity and drinking water. Still, most people do not have the money to return to their villages. They remain in poverty-stricken urban neighborhoods with substandard housing and poor sanitation.

History and Government

History and government in Central America and the Caribbean have been influenced by indigenous cultures, **colonialism**, slavery, and struggles for freedom.

European Conquests

The arrival of Christopher Columbus in the West Indies triggered the conquest and colonization of Central America and the Caribbean islands. He explored and colonized the Caribbean islands from 1492 to 1504. The first permanent European settlement was established on the island of Hispaniola in 1493. Large numbers of Spanish settlers seeking gold followed.

The Spanish brutally conquered the Native Americans and forced them to work as slaves in gold mines and on plantations. By 1600, hard labor, starvation, and European diseases nearly destroyed all the natives. To meet the labor shortage, European colonists imported Africans as slaves.

Columbian Exchange

Columbus's arrival began one of the most significant events in world history, the **Columbian exchange**. As Europeans arrived to claim lands for Spain, Portugal, France, and Britain, they introduced food plants like wheat, oats, rice, sugarcane, coffee, and grapes as well as domesticated animals like cattle, horses, sheep, goats, pigs, and fowl. Europeans also brought the natives new diseases like smallpox, influenza, measles, yellow fever, and malaria, for which the natives had not developed immunity.

Native plants, animals, and diseases were taken back to Europe. Potatoes, beans, maize, tobacco, tomatoes, and cacao became parts of the European diet. Europeans also took llamas, turkeys, alpaca, and syphilis back across the Atlantic Ocean. Diseases were unintentionally transmitted.

Panama Canal

Vasco Nuñez de Balboa explored the isthmus now known as Panama, and discovered the Pacific Ocean. Almost four centuries later, business interests and politicians realized it would be profitable to build a shortcut between the oceans. In 1904, construction began on the Panama Canal. Nearly 75,000 laborers from around the world built one of the engineering wonders of the world. Today, the canal is still an important route for trade.

DID YOU KNOW?

When the Colombian legislature would not approve of foreign ownership of the canal, U.S. President Teddy Roosevelt triggered a revolution in what would become Panama to seize the land on which the canal would be located.

Gaining Independence

Native Americans and African slaves yearned for freedom by the late 1700s. In the 1790s, Francois Toussaint L'Ouverture, a soldier born to enslaved parents, led enslaved Africans to fight against Spanish and French colonial powers in Haiti. He died in 1802, but his efforts helped pave the way for the declaration of Haiti's independence on January 1, 1804.

Besides Haiti, most Caribbean countries were the last territories in the region to achieve independence. Cuba won its independence from Spain in 1898, but remained under U.S. protection until 1902. British-ruled islands like Jamaica and Barbados did not win independence until well into the 1900s. Some Caribbean islands still remain under foreign control. Puerto Rico and some of the Virgin Islands have political links to the United States. Spain ruled Central America until the nineteenth century.

Struggles for independence started a period of economic and political instability. During the 1800s, leaders sought to construct political institutions and prosperous economies. In 1823, independent provinces formed a federation called the United Provinces of Central America. Powerful elites opposed the union, so the United Provinces separated into the countries of Guatemala, El Salvador, Honduras, Nicaragua, and Costa Rica.

Movements for Change

Many Central American and Caribbean countries experienced dramatic political, social, and economic changes throughout the 1900s. Railroad building and the formation of industries brought new wealth to the upper classes. Most people, especially those in rural areas, saw little progress. The demands of the poor for reform were ignored.

In Cuba, Fidel Castro led a successful guerilla revolution and set up a **communist** state. Communism remains entrenched in Cuba, but there have been some reforms in the 2000s.

Military dictatorships gave way to democratically elected governments in several other Latin American countries in the 1990s. Today, many countries in Central America and the Caribbean struggle to end corrupt politics and violence and bring economic benefits to all citizens. In the last few years, many people in Latin America voted and expressed their desire for change.

Language and Religion

Spanish is the main language in Central America. In the Caribbean, the European languages of English, Spanish, French, and Dutch are spoken. Each country has its own dialects, or forms of the language, unique to a particular place or group.

Millions of people speak Native American languages. Many are bilingual. Haitian Creole has a mainly French vocabulary with some African and Spanish words.

DID YOU KNOW?

Others speak numerous forms of **patois**, dialects that combine indigenous, European, African, and Asian languages.

The majority of people in Central America are Roman Catholic. In the Caribbean, most living on Spanish- or French-speaking islands are Roman Catholic, but various Protestant denominations are found in the English-speaking areas. Islam and Hinduism are also found in the region, as are numerous traditional Native American and African religions, which are often mixed with Christianity and other faiths.

DID YOU KNOW?

Santeria in Cuba and voodoo in Haiti and the Dominican Republic are examples of mixed religions based on African rituals.

Education and Healthcare

The quality of education varies by country as well as in rural and urban areas. Children are required to finish elementary school, but few do because of long distances to school and lack of money for supplies and clothing.

Healthcare is linked to standards of living. Countries with a developed welfare system have higher life expectancies and standards of living. Countries with less developed economies have little money to spend on healthcare, so disease and malnutrition are common and life expectancy low.

Arts

Native Americans produced the earliest forms of artwork, including wood carvings, pottery, metalworks, and weaving. The work of pre–Columbian artisans was often as sophisticated as anything made by hand today. Hand-woven textiles made in villages today reflect ancient Mayan symbols and weaving techniques.

The music of Central America and the Caribbean, like salsa or reggae, combines Native American, European, and African styles and dances. Many styles evolved out of the traditional music of indigenous wind and percussion instruments, European string and brass instruments, and African drums, rhythms, and dances.

Family Life

Throughout Central America, the extended family is the basic unit of society. The importance of family in a community is a factor determining one's social class. In the Caribbean, family structure is often **matriarchal**, ruled by a woman such as a mother, grandmother, or aunt. This family structure type is representative of West Africa, where many Caribbean people have roots.

Sports and Leisure

Baseball, basketball, cricket, and volleyball have large followings in the Caribbean. American sailors in Cuba taught baseball to the Cubans over a hundred years ago. Cubans took the game with them wherever they migrated. In the Dominican Republic, baseball has become the national pastime. Soccer is popular in Central America.

Cash Crops

The fertile highlands of Mexico and Guatemala have enabled them to become two of the world's leading coffee producers. The tropical climate of Cuba puts it among the world's leading producers of sugarcane.

Countries assume great risk by only growing one or two export products. Droughts, floods, or volcanic eruptions can destroy a country's cash crops and ruin the economy. In 1998, Hurricane Mitch destroyed 90% of Honduras's banana crop. In 2004, Hurricane Ivan caused severe damage to the Caribbean, destroying Grenada's cash crop nutmeg.

CAFTA

In 2005, the United States joined six Central American countries in the Central American Free Trade Agreement (CAFTA). It lowered trade barriers between the United States, Costa Rica, El Salvador, Guatemala, Honduras, Nicaragua, and the Dominican Republic. Critics are concerned about American job losses and the exploitation of lower paid workers, especially in the sugar and textile industries.

Practice

1. People who are part Native and part European are called ___________.

2. In Central America, most people live in the ___________ that run along the Pacific coast.

3. ___________ were the first slaves in the Caribbean.

4. The exchange of food, plants, and diseases by Natives and Europeans in the 1500s was called the ___________.

5. Most people in Central America speak ___________.

6. ___________ is the national pastime in the Dominican Republic.

7. The ___________ cut the distance in shipping between the Atlantic and Pacific oceans.

8. ___________ was the first Caribbean country to free itself from slavery.

9. Santeria in Cuba and voodoo in Haiti and the Dominican Republic are examples of ___________ religions, or syncretism.

10. The ___________ family is important in Central America and the Caribbean.

11. Latin America is susceptible to natural disasters such as ___________ and volcanic eruptions.

12. The United States joined six Central American countries in lowering trade barriers through ___________.

Answers

1. Mestizo
2. Highlands
3. Native, or indigenous, peoples
4. Columbian exchange
5. Spanish
6. Baseball
7. Panama Canal
8. Haiti
9. Mixed
10. Extended
11. Hurricanes
12. CAFTA

LESSON 10

LATIN AMERICA—HUMAN GEOGRAPHY (PART II)

LESSON SUMMARY

This lesson examines the human geography, or population, culture, language, religion, economy, education, healthcare, arts, and family life of South America and the environmental challenges facing it. South America has been influenced by Native American and European cultures, migration, physical geography, and urbanization.

Latin America is very fond of the word "hope." We like to be called the "continent of hope." Candidates for deputy, senator, president, call themselves "candidates of hope." This hope is really something like a promise of heaven, an IOU whose payment is always being put off. It is put off until the next legislative campaign, until next year, until the next century.

—Pablo Neruda

Population Patterns

South America's population has been shaped by ethnic diversity, physical geography, migration, and urban growth.

People

South America has an ethnically diverse population. Indigenous cultures inhabit parts of the region, particularly in rural and remote areas. Hundreds of indigenous groups live in the Andes region of Ecuador, Bolivia, and Peru.

The Spanish and Portuguese were the first Europeans to settle in South America. Africans were brought as slave labor. After the South American countries gained their independence, other European groups like the French, Dutch, Italians, and Germans arrived, as well as immigrants from Asia. Half of the population of Guyana is of South Asian or Southeast Asian ancestry. Many ethnically Chinese people call Peru their home. Many people of Japanese descent call Brazil, Argentina, and Peru home.

Density and Distribution

In South America, the challenges of physical geography are increased by a high rate of population growth. Human settlement is difficult because the interior of South America is made up of rain forests, deserts, and mountains. Most people tend to live on the populated rim of South America. The coasts provide favorable climates, fertile land, and easy access to transportation systems. To draw people in from the overpopulated coast, the Brazilian government in 1960 moved the capital from coastal Rio de Janeiro to Brasilia, a planned city in the country's interior.

South American countries have large land areas, so their population densities are low. Ecuador is the most densely populated country in South America, with 135 people per square mile.

DID YOU KNOW?

Brazil has a population of 203 million spread out over 3.3 million square miles, or 61 people per square mile.

Many people have left South America to escape civil war violence in search of better wages and living conditions. However, migration from South America to the United States is low compared to the high levels of immigration from Mexico to the United States. Some countries like Colombia, Ecuador, and Guyana have experienced what some call a **brain drain**, or loss of highly educated and skilled workers to other countries.

The population of South America has become predominately urban because of migration. About 80% of the subregion's population is urban. In Argentina and Uruguay, urbanization has been the result of foreign immigration. In most South American countries, urbanization has been the result of internal migration.

Urban Challenges

Brazil's São Paulo and Rio de Janeiro and Argentina's Buenos Aires rank among the world's largest urban areas in terms of population. These megacities have extreme divides between rich and poor. Some demographers estimate that in São Paulo, 20% of the city is living in favelas, or slums on the outskirts. The challenges for São Paulo and other megacities are housing, employment, maintaining infrastructure, crime, and traffic.

History and Government

Indigenous civilizations, colonization, independence, and authoritarian rule have influenced South America's history and government.

Early Cultures

The Moche, Mapuche, and Aymara developed agricultural societies in South America well before the Inca. The Inca developed a highly developed civilization in the Andes. The Inca Empire stretched at its height in the late 1400s and early 1500s from Ecuador to central Chile. The empire was called Tawantsinyu, meaning "land of the four quarters," which met in the city of Cuzco, now in Peru.

The Nazca people, who lived in Peru's southern valleys from around 300 B.C. to A.D. 800, produced remarkable Nazca Lines in the earth. Located in the Nazca Desert, these shallow glyphs in the shapes of animals span up to 660 feet wide. Some speculate that they were used to predict seasons for farming or were used for extraterrestrial communication, a claim often applied to the mountaintop development of Macchu Picchu as well.

The Inca maintained a central government headed by an emperor. In this society, the emperor, head priest, and army commander held complete authority over all other classes. The farmers, artisans, and laborers made up the lower classes.

The Inca were skilled engineers. Inca pyramids are infinitely more complicated and precisely cut than Egyptian pyramids. No one knows how they could have done the complicated and mathematically precise cuts to make the unusually shaped stones that they then fit together. The Inca laid 25,000 miles of footpaths or roads, a network about half as large as today's interstate highway system in the United States. Their roads crossed mountain passes and penetrated forests.

Inca farmers cut terraces into the slopes of the Andes to make irrigation systems for their crops of cacao they used for headaches and fatigue. They kept no written language, so most of what we know about them comes from oral history and storytelling. The Inca used quipu, a series of knotted cords of various colors and lengths, to keep financial and historical records.

The Inca were quite wealthy since they had vast mineral resources of gold and silver. This great wealth brought the Spanish conquistadors to Peru. The Inca road network made it easier for the Spanish to subdue the empire.

European Conquests

The Inca were defeated by diseases, civil war, and the Spanish. The Spanish conquistadors expanded from Peru into Colombia, Argentina, and Chile. The Portuguese settled on the coast of Brazil. The British, French, and Dutch settled across the northern part of South America.

The Spanish and Portuguese conquerors set up highly structured political systems, such as the viceroyalties of New Granada, Peru, La Plata, and Brazil. The Europeans did this wherever they established settlements or colonies. The Roman Catholic Church was the unifying institution and intermediary in South America.

European colonies in South America became sources of great wealth for their empires. Some Spanish settlers mined for silver and gold, and such a great quantity of the precious metals was brought back that the Spanish economy faced inflation for 100 years. The Portuguese discovered not only gold and silver but also Brazil wood, used to make red dye. On Spanish, Portuguese, and Dutch plantations, sugar, coffee, and cotton were harvested and processed for export to Europe using first Native then African slave labor.

The Native American populations were annihilated in South America by European diseases and the hardships of colonial plantation slavery. The decimation of the Native populations resulted in the European importation of more African slaves to meet their labor shortages. In many colonies, like the Dutch colony of Suriname, the number of slaves held a majority until the mid-nineteenth century.

Independence

The American, French, Mexican, and Caribbean revolutions inspired South Americans to fight for their independence from European colonial rule. By the mid-1800s, most of South America achieved this goal under the leadership of Simón Bolívar in Venezuela and José de San Martin in Argentina. Brazil achieved independence without violence. Suriname did not obtain independence until 1975. French Guiana remains part of France, although most of the population wants independence.

The post-colonial period was unstable politically and economically for the new independent countries of South America. They lacked a tradition of self-government. Even though these countries drafted and approved constitutions, power remained in the hands of the wealthy elite. Caudillos, or authoritarian political leaders, frequently seized power illegally by force with the help of the military.

Movements for Change

In some South American countries, dictatorships were followed by democratically elected governments. These countries now struggle to end political corruption and violence, shrink the gap between rich and poor, create jobs, and strengthen indigenous rights.

The struggle for **democracy** in South America continues. In Colombia, the government has been fighting paramilitary and insurgent groups for 40 years. They are not strong enough to topple the government, but they do control some of the rural areas.

Ecuador has had civilian control of the government since 1979, but has had seven presidents since 1996. Peru was ruled by the military for 12 years until 1980. Until president Fujimori instituted economic reforms in the 1990s, the government faced civil insurrections. Although he left office in 2000, his government is still criticized for corruption. In Bolivia, the government was overthrown 200 times from 1825 to 1982. Since 1982, Bolivia has remained democratic.

Demands for change have been made through the ballot box. In 2006, Chile elected its first female president, Michelle Bachelet. Bolivians elected Evo Morales, an Aymara Indian, as their first Native American president.

Culture

The culture of South America has been influenced by the beliefs, traditions, and arts of indigenous peoples, Europeans, and Africans. Diverse cultural influences are found in cities and villages.

Language and Religion

In different parts of South America, people speak Spanish, Portuguese, French, or Dutch. Many people are bilingual. During colonization, some European languages mixed with Native American languages to form completely new languages. Many Native American languages are still spoken in South America.

Most South Americans are Roman Catholic. Tens of millions of people practice mixed religions like Macumba, a word with negative connotations to some, and Candomblé, which combine West African religions with Roman Catholicism. Protestant Christianity, Hinduism, Buddhism, Shintoism, Judaism, Eastern Orthodox Christianity, and Islam are also practiced in South America.

Education and Healthcare

Education varies throughout South America. Many countries have devoted more funds to public schools. Literacy rates have thus risen steadily. Some public universities provide higher education at little or no cost to students. In other countries, education is a luxury, and many poor children drop out of school to support their families.

In countries with stable economies and higher standards of living, people have access to healthcare and live longer, healthier lives. In the remote or rural areas of South America, health concerns arising from poverty, lack of sanitation, infectious diseases, and malnutrition remain. This concern is also strong in the favelas, or **shantytowns**, that exist in the slums on the outskirts of large cities where millions of people live in dirty, overcrowded conditions.

Arts

Native American arts survive in many forms. The massive buildings of the ancient Inca at Cuzco and high up in Macchu Picchu, "the hitching post of the gods," baffle our contemporary understanding of masonry. Traditional crafts like weaving, ceramics, and metalworking have been passed down through the generations and are still practiced today.

Ancient music still influences the modern in South America. The panpipe, a common pre–Colombian musical instrument, was used widely in the Andes. Native American, African, and European influences have combined to create unique styles like the Brazilian samba and Argentine tango. African slaves developed a Brazilian martial arts form disguised as dance called capoeira.

In colonial times, Spanish art forms influenced South American painting and architecture. Spanish- and Portuguese-style Catholic churches still remain. Native American and African artists often added color to South American architecture. Brazilian architect Oscar Niemeyer is known for his design of modern buildings in Brasilia.

Many South American writers have won international renown. Colombian novelist Gabríel García Márquez mixes everyday reality with fantasy. Chilean poets Gabriela Mistral and Pablo Neruda have won Nobel Prizes for Literature.

Family Life and Leisure

The family is more likely to be nuclear in a middle- or upper-class South American family, but loyalty and responsibility toward the extended family remain strong. The *compadre* relationship (the bond between the parents and godparents of a child) is still valued, but changes brought by urban society have lessened its importance in some places.

People love soccer in South America. Polo, auto racing, tennis, boxing, and basketball are also popular sports. Family visits, patriotic events, religious holidays, and festivals engage people in social life and leisure time. Carnival is celebrated the week before Lent, a 40-day period of fasting and prayer for Roman Catholics before Easter. Rio de Janeiro has one of the largest annual Carnival celebrations in the world.

Economy

South American countries face the challenge of developing and diversifying their economies. Still, there are large disparities between the rich and poor in major cities.

Agriculture

Although about 80% of South America's people live in cities, agricultural exports like bananas, coffee, and sugarcane bring in a major portion of their national incomes.

Land has been unevenly distributed in South America for 400 years. Wealthy family or corporate agricultural estates called **latifundia** are worked by rural farmers called **campesinos**. Today, latifundias are highly commercial and mechanized. They yield high returns with very little investment in labor. Smaller farms are called **minifundia**. These small plots of land are farmed intensively by the campesinos to feed their families. The campesinos usually do not own minifundia land—wealthy landowners, partnerships, or the government do.

The latifundia and minifundia systems are dissolving. As the latifundia mechanize, farmers are leaving to look for city jobs. Governments are passing laws to distribute the land more fairly. Many campesinos have formed agricultural cooperatives by combining minifundia into large, jointly owned farms. The legacy of the campesinos is hard to break and so most remain poor.

Cash Crops and Livestock

The fertile highlands of Brazil and Colombia have allowed them to become two of the world's leading coffee producers. The tropical climates and fertile soils of Brazil put it among the world's leading producers of sugarcane. One of the largest cash crops in Brazil, soybeans, is used to feed cattle. Cash crops benefit commercial farmers most. Some South American countries raise cattle on large ranches for export.

Industry

Industrial growth has been limited in South America by physical features like the high Andes and lush tropical rain forest of the Amazon, which make it difficult to access natural resources. Foreigners brought new technology to South America, but they drained the local resources and profits. Investors are also wary of investing due to political instability.

Some countries are overcoming these limitations by combining necessary resources with stable governments and active business communities. Brazil, as an example, emerged from financial crises in the 1990s to expand global trade and their overall economies.

Trade and Interdependence

Many South American countries developed their economies by promoting trade and decreasing foreign debts. They trade to obtain the natural resources, manufactured goods, and foods they do not have or cannot produce. Trade within the region and the world has begun to increase.

Managing Resources

South America is working to protect the environment while facing rapid urbanization and growing human needs. Deforestation, or the clearing and destruction of forests, is occurring in South America and other places in the world. One strategic solution is **sustainable development**—technological and economic growth without depleting the human and natural resources of an area.

Farms versus Forests

Logging, farming, and ranching threaten the survival of the Amazon rain forest ecosystem. Thirty percent of people in South America work in agriculture. Expanding livestock pastures, so important to the economy of South America, increases deforestation. Farmers clear large rain forest areas to grow cash crops. Deforestation is particularly severe in the Brazilian rain forests where multinational agricultural companies sponsor large-scale conversion of rain forests into large plantations.

In the Amazon basin, slash and burn farming is practiced. Plants are cut down and trees are stripped of bark. It is all dried and set on fire. The ash provides nutrients for the soil. Rains **leach** the benefits away and the soil is left infertile. Crop yields decline and farmers clear new parts of the forest. Latifundia and large corporations continue to expand their soybean fields to meet world demand.

DID YOU KNOW?

In the commercial logging business, two-thirds of the timber cut down in the Amazon is not ever used.

Biodiversity at Risk

Rain forests provide sanctuary to 50% of all animal and plant species on Earth. Deforestation threatens these ecosystems. Over 20% of the Amazon rain forest has already been destroyed. The Brazilian Atlantic forest is also threatened and scientists are creating corridors of vegetation to connect what remains of it.

Deforestation threatens the resources in the Amazon. Medicines are made from rain forest plants and organisms; some may even cure cancer. Loss of the rain forest also increases the amount of carbon dioxide in the air, resulting in increased global warming, climate change, and increased ocean levels.

Transportation

Roads and railroads in South America cross rugged mountains, dense rain forests, and arid deserts. The Pan-American Highway runs from Southern Chile to northern Mexico, and extends through the United States into Alaska. It links more than a dozen Latin American capital cities. A trans-Andean highway links cities in Chile and Argentina. The Trans-Amazonian Highway built by Brazil cuts across the Amazon rain forest. Peru and Brazil are building a transoceanic highway to link the Amazon River in Brazil with Peru's ports on the Pacific Ocean. This highway will take products from Brazil and Peru to global markets and increase trade between the two countries.

Brazil and Argentina have well-developed rail systems despite the physical barriers that restrict people from using them. Railroads have fallen into disrepair. The inland waterways of the Amazon River and Paraná-Paraguay Rivers are important. If air travel becomes less costly, it will help overcome geographic barriers. All South American capitals receive domestic and international flights. Remote locations are served by private and military landing strips.

Disputed Borders

South American countries have fought over the past 150 years for strategic location or the ownership of natural resources. Border wars divert resources away from development. In 1998, Peru and Ecuador stopped their 60-year border dispute.

Human Impact

Rapid urbanization and industrial growth have placed tremendous stress on available natural resources in South America.

Rapid Urban Growth

When population growth far exceeds available resources, it is called rapid urbanization. Rural workers who migrate to cities in search of work often end up in slums or shantytowns without work. These shantytowns often lack running water and underground sewage. They are unsanitary and disease-ridden.

Air pollution is a problem in cities without adequate clean-air regulations. Exhaust gases from clogged streets pollute the air; industrial smokestacks also belch toxins into the air.

Governments, international agencies, and grassroots organizations are doing what they can. Groups that advocate for the homeless in Santiago, Chile, turned abandoned buildings into affordable housing.

Industrial Pollution

Industrial growth has increased due to free trade agreements and the expansion of multinational corporations. Environmental laws have not reduced the risks of increased pollution.

DID YOU KNOW?

Runoff from chemical fertilizers and pesticides used by commercial farms can cross borders and damage people's health.

Disaster Preparedness

South America's physical geography makes it susceptible to natural disasters. In 2005, there were many devastating hurricanes. Latin American governments are now investing in emergency preparedness. Satellite imaging and computer modeling technologies are now being purchased to forecast the direction and severity of hurricanes.

Scientists are also gathering information about volcanic eruptions that started on the Caribbean island of Montserrat in 1995 and left two-thirds of the island uninhabitable. Because this volcano is similar to others on different continents, the information will be used to prepare for future volcanic eruptions.

Practice

Insert T or F to indicate whether the statement is true or false, based on what you've read.

1. _____ In South America, caudillos frequently seized power illegally by force with the help of the military.

2. _____ 80% of South America's people live in rural areas.

3. _____ Slash and burn practices are excellent for environmental conservation.

4. _____ São Paulo and Rio de Janeiro have rapidly urbanized.

5. _____ Replanting rain forest trees is a form of sustainable development.

6. _____ Deforestation of the Amazon is harmful to the planet.

7. _____ Colombia's cash crop is coffee.

8. _____ Population density is highest in the interior of South America.

Answers

1. T
2. F
3. F
4. T
5. T
6. T
7. T
8. F

RUSSIA—PHYSICAL GEOGRAPHY

LESSON SUMMARY

Spanning Europe and Asia, Russia is made up of towering mountains, active volcanoes, lowlands, plains, and vast tundra. Although the Soviet Union broke apart into fifteen republics in 1991, Russia still has the largest land area of all the republics. Russia's location in the far northern latitudes, its interconnected mountain ranges, and its large river systems still influence human settlement and activities. Russia's location deep within the Eurasian landmass also affects the region's climate and vegetation.

I came to know the world's largest boreal forest from the taiga to beyond the Arctic Circle. . . . I am partial to the light of the north woods—slanting rays that in the warmer months cast long evening shadows and suffuse the landscape with a crystalline glow.

—Fen Montaigne, "The Great Northern Forest," *National Geographic* (June 2002)

Landforms

Russia is the largest country in the world. Russia is more than a third larger than the United States, but has little variety in landforms. Russia is generally flat. Lowland plains cover the west, plateaus cover the rest. Russia is divided and bordered by mountain ranges, tundra, subarctic forests, and wide rivers and seas.

Mountains and Plateaus

Russia is commonly referred to in terms of five land regions: the European Plain, the Ural Mountains, the West Siberian Plain, the Central Siberian Plateau, and the East Siberian Uplands. Each region differs in soil, vegetation, and human activity.

European Plain

About 75% of the Russian population lives on the Northern European Plain, or Russian Plain, that sweeps across western and central Europe. Russia's most populous cities, Moscow and St. Petersburg, were built on the Northern European Plain.

The northern part of the plain is poorly drained, so there are swamps and marshes. The southern part is rich in the **chernozem** soil that supports most of Russia's agricultural production of wheat, barley, and other crops.

Caucasus Mountains

The Caucasus Mountains of southwest Russia stand between the Black Sea and the Caspian Sea on the European Plain. Mount Elbrus, an extinct volcano, is Russia's highest peak at 18,510 feet. This subregion has a moderate climate.

Ural Mountains

The Ural Mountains form a **natural boundary** between European Russia and Asian Russia. The Urals extend roughly 1,500 miles, with the tallest peaks reaching over 5,000 feet. They are rich in iron ore and mineral fuels like oil and natural gas. The Urals are home to industrial development.

West Siberian Plain

The Ural Mountains divide the European Plain from the West Siberian Plain. The poorly drained West Siberian lowlands have many swamps and marshes.

The 1 million square miles of plains and plateaus called Siberia stretch all the way from the east side of the Ural Mountains to the Pacific Ocean, and from the Arctic Ocean in the north to the Central Asian grasslands in the south.

Central Siberian Plateau

The Central Siberian Plateau forms a natural boundary between Russia and China. The plateau's 1,600 to 2,300 foot peaks formed swiftly moving rivers and canyons that can still be found there.

East Siberian Uplands

In the east, various mountain ranges and basins extend from Siberia to the Pacific Ocean. The Kamchatka Peninsula has more than 100 volcanoes, of which more than 20 are active.

Water Systems

Coasts, Seas, and Lakes

Russia's 23,400 miles of continuous coastline is the longest in the world. Russia's coastlines reach the Arctic and Pacific Oceans and the Baltic Sea, Black Sea, and Caspian Sea.

The Black Sea gives Russia's warm water access to the Aegean and Mediterranean Seas through the Turkish-controlled Bosporus, the Sea of Marmara, and the Dardanelles. The world's largest inland body of water, the Caspian Sea, is actually a saltwater lake. Rivers flow into the Caspian Sea, but it does not have any outlet to the ocean.

The world's oldest and deepest freshwater lake is Lake Baikal. Located in southern Siberia, Lake Baikal is 400 miles long, 40 miles wide, and a mile deep. It has more water by volume than any lake in the world. Lake Baikal contains 20% of Earth's supply of freshwater.

DID YOU KNOW?

More than 25 million years old, Lake Baikal is fed by over 300 rivers and streams and is home to about 50 fish species.

Rivers

Some of the world's longest rivers run through Russia. Russia's large river systems are essential for irrigation, transportation, electrical power, and industries like fishing.

TIP

Only 25% of Russia's people live in Siberia, which provides 84% of the country's water.

Volga River

Western Russia's Volga River is vital to Russia. *Matushka*, or "Mother," Volga drains most of the eastern part of Russia's Northern European Plain. The Volga flows through areas of temperate grasslands and mixed forests.

The Volga provides important transportation links between Moscow and the Caspian Sea and through the Volga-Don Canal to the Sea of Azvov and the Black Sea.

Melting snow gives the Volga River 33% of Russia's usable water. Its waters are used for drinking, irrigation, and hydroelectric power. Most of this water is returned with waste to the Volga River.

Siberian Rivers

The Siberian Rivers include the Ob, Irtysh, Yenesey, and Lena. These rivers make up one of the world's largest river systems. They flow north to the Arctic Ocean. In the north, where rivers are blocked by ice, swamps and marshes form. The 1,000 mile long Amur River makes up the border between Russia and China. The Amur River drains in the east.

DID YOU KNOW?

Warm monsoon winds from the southeast make the Amur River Valley one of Siberia's major food-producing areas.

Natural Resources

Russia's abundant natural resources are often located in remote, inhospitable, and inaccessible parts of the country.

Minerals and Energy

Russia is one of the world's leading producers of fossil fuels. Most of Russia's petroleum and coal deposits can be found in Siberia and the Ural Mountains. Russia has ample natural gas deposits in northern Siberia. Russia's rivers make Russia a leader in hydroelectric power

DID YOU KNOW?

Russia leads the world in nickel production and is among the top three in aluminum, platinum, and gemstone production.

Soil and Forests

Approximately 20% of the world's remaining forest lands are in Russia. Most of Russia's forests lie in eastern Siberia. Russia's **boreal forests** provide much of the world's timber, particularly pine, spruce, cedar, and fir.

Commercial logging and wildfires have reduced Russian forests by 40 million acres a year, which is more than the Amazon River basin. These forests regulate climate and filter out billions of tons of carbon dioxide and other gases during photosynthesis. The carbon is stored in trees, roots, and soil.

DID YOU KNOW?

Only the Amazonian tropical rain forest returns more oxygen to the atmosphere than Russia's northern boreal forests.

Fishing Industry

The fishing industry is important to the Russian economy and diet. Salmon are caught in the Pacific Ocean. Cod, herring, and halibut, are hauled in from the Arctic Ocean. The supply of world-famous caviar, processed and salted fish eggs, has declined, though, because dams on the Volga River have interrupted the sturgeon migrations needed for the finest caviar. Global demand for caviar is usually met now through illegal fishing.

Practice

Fill in the blank with the correct word from the information in the preceding paragraphs.

1. The ___________ Mountains in Russia separate Europe from Asia.
2. The Russian ___________ forests are losing 40 million acres a year, more than the Amazon loses.
3. ___________ provides 84% of Russia's water.
4. About 75% of the Russian population lives on the ___________ Plain.
5. The peaks of the Central ___________ Plateau form a natural boundary between Russia and China.
6. The world's oldest and deepest freshwater lake is Russia's Lake ___________.
7. ___________ lands dominate Central Asia.
8. ___________ is a major industry in the Arctic and Pacific oceans.
9. Russia has the longest continuous ___________ in the world.
10. The world's largest inland body of water, the ___________ Sea, is actually a saltwater lake.

Climate and Vegetation

Climate shapes settlement patterns. Russia's location in the northern latitudes means long, cold winters. People must adapt their jobs, transportation, food, water, heating, clothing, and plumbing to the Russian cold. Winters are especially harsh in Siberia, so people tend to live in the west.

High Latitude Regions

Russia's location in the high latitudes of the Eurasian landmass results in a harsh climate with long, cold winters and short, cool summers. Seasonal temperatures can vary greatly.

In eastern Russia's Yakutsk, January temperatures often fall below –33°F while July temperatures average 64°F. Eastern Siberia has the coldest winter temperatures. January temperatures at Verkhoyansk, the "cold pole" of the world at 68° N latitude, have fallen to –90°F.

In western Russia, warmer air from the Atlantic Ocean moderates temperatures. Because most of the Eurasian landmass does not receive ocean winds, there is little precipitation. The interior also faces extreme temperature variations. This is known as **continentality**.

Tundra

The tundra is the vast treeless northern plain almost entirely north of the Arctic Circle at 66.5° N. The average annual temperature of the tundra is below freezing. The tundra covers about 10% of Russia. The thin acidic soil above the permafrost can only grow mosses, lichens, algae, and dwarf shrubs.

DID YOU KNOW?

For many weeks before and after December 22, the sky stays dark in the tundra. In the summer, there is continuous sunlight in the tundra for several weeks.

Subarctic

South of the tundra is Russia's largest climate region, the subarctic. Some of the world's coldest temperatures have been recorded in the subarctic. Snow covers the ground 120 to 250 days a year. The **taiga**, a boreal forest belt about the size of the United States, covers 40% of western Russia. The taiga is the world's largest coniferous forest, containing about 50% of the world's softwood timber.

Midlatitude Regions

Russia's midlatitude regions have more moderate climates. The midlatitudes have milder winters and warmer summers. The midlatitudes support most of the country's population and agricultural production.

Humid Continental

Most of the North European Plain has a humid continental climate. Moscow lies in a humid continental region. Moscow's temperatures range from 9 to 14°F in January, and 66 to 99°F in July.

In humid continental areas, there are mixed coniferous-deciduous forests. Soils in these areas use farming methods and fertilizers productively. In the southern portion of the humid continental climate, mixed forests turn into temperate grasslands.

DID YOU KNOW?

The rich chernozem soil, with its high concentration of humus and phosphoric acids, is fit for production of crops like wheat, sugar beets, and sunflowers.

DID YOU KNOW?

When Napoleon arrived back in Poland from Russia, only 40,000 of his 600,000 men were still alive after facing the harsh Russian winter.

In World War II, when Hitler and his armies arrived near Moscow in December of 1941, frigid –40°F temperatures paralyzed his tanks, mechanized vehicles, artillery, and aircrafts. It was one of the coldest winters of the century. Russia's harsh winter was an important factor leading to the Germans' retreat.

Steppe

The Russian steppe lies north of the Caucasus Mountains between the Black Sea and the Caspian Sea. There is also a band of steppe along the Russian border with Kazakhstan.

The steppe has dry summers and long, cold, dry winters with swirling winds and whirling snow. Plants flourish in the steppe's organic rich chernozem soil. The steppe's sea of grass stretches to the horizon. Sunflowers, mints, and beans grow in the steppe, but overgrazing animals and foreign plants have damaged the steppe ecosystem. As nonnative species crowd out native grasses, soil fertility declines.

Practice

Insert T or F to indicate whether the statement is true or false, based on what you've read.

11. ____ Sunflowers, mints, and beans flourish in the tundra's organic rich chernozem soil.

12. ____ Western Russia enjoys more moderate temperatures because of warm winds from the Atlantic Ocean.

13. ____ Moscow experiences a humid subarctic climate in the midlatitudes.

14. ____ The Russian tundra faces weeks of continuous sunlight and weeks of constant darkness every year.

15. ____ Russia's location in the northern latitudes means long, cold winters.

16. ____ In the steppe, there are mixed coniferous-deciduous forests.

17. ____ The midlatitudes support most of the country's population and agricultural production.

18. ____ Seasonal temperatures vary little in Russia's high latitudes.

19. ____ Most people live on the eastern side of Siberia where abundant mineral resources are quite accessible.

20. ____ The taiga is the world's largest coniferous forest, containing about 50% of the world's softwood timber.

Answers

1. Ural
2. Boreal
3. Siberia
4. North European or Russian
5. Siberian
6. Baikal
7. Grass
8. Fishing
9. Coastline
10. Caspian
11. F
12. T
13. F
14. T
15. T
16. F
17. T
18. F
19. F
20. T

LESSON

12 RUSSIA—HUMAN GEOGRAPHY

LESSON SUMMARY

Russia's population includes more than 150 ethnic groups, including Slavic, Turkic, and Caucasian peoples, speaking a total of more than 100 languages. In this chapter, we follow Russian history from the Slavic settlements along the waterways of the North European Plain, through the Czars and Communist rule, to the environmental challenges facing Russia today.

Anyone who doesn't regret the passing of the Soviet Union has no heart. Anyone who wants it restored has no brains.

—Vladimir Putin, President of Russia

Population and Culture

Almost two-thirds of the people in Russia are ethnic Russians with a common language, history, and tradition of strong central government.

For most of the twentieth century, Russia had a government-controlled economy led by Communist Party dictators. When the Soviet Union dissolved in the early 1990s, Russia adopted democracy and a market economy.

Population Patterns

Ethnic groups, migrations, and invasions have shaped population patterns in Russia. Each distinct ethnic group has a common ancestry, language, religion, and customs.

Today, low birth rates mean Russia's population is aging and straining national resources, just like in the United States.

People

Russia's historical roots go back thousands of years and include numerous ethnic groups. As Russia grew over the centuries from a territory to an empire stretching from Europe to the Pacific Ocean, many non-Russian ethnic groups came under its control. During the Soviet era of 1922-1991, Russia was part of the Union of Soviet Socialist Republics (USSR). Regional political boundaries reflected ethnic group nationalities.

Many of the larger republics declared independence after 1991. Russia remains diverse. There are 32 ethnic groups within Russia with their own republics or administrative territories.

DID YOU KNOW?

About 80% of Russia's people are Ethnic Russians. Ethnic Russians are the largest ethnic group in Russia, and one of the largest in the world.

Slavs

In the 600s, Slav farmers, hunters, and fishers settled along the waterways of the North European Plain. In the 1200s, Slavs fled invading Mongol hordes and settled by the Moskva River. The Slav settlements formed the territory of Muscovy. Muscovy's city center of Moscow was surrounded by plentiful farm and hunting grounds, and it was linked by rivers to major trade routes.

In the 1400s, Ivan III, "the Great," brought many Slav territories under his control. In Moscow, he built the Kremlin fortress and built churches and palaces.

DID YOU KNOW?

One of the most iconic features of Russian architecture is the onion dome. Some say that onion domes represent the vault of heaven, while others believe the shape is a practical way to prevent the accumulation of snow.

Ethnic groups referred to as Slavs are considerably diverse both culturally and in appearance. Russians are included, as well as Poles, Serbs, Ukrainians, and other Eastern Europeans.

Caucasians

Caucasian peoples such as the Chechens, Dagestanis, and Ingushetians live in the Caucasus region of southwest Russia.

Turks

Turkic peoples like the Tatars, Chuvash, Bashkirs, and Sakha live in Central Asia, Eastern Europe, the Mediterranean, and the Middle East. In Russia, they live mostly in the Caucasus and Middle Volga areas. The western republic of Tatarstan has been ruled by Russia since the 1550s but enjoys limited **sovereignty,** or self-rule.

The Sakha are a mixture of local groups and formerly seminomadic Turks. The Sakha settled along the Lena River and in southern Siberia and expanded into northeast Russia.

Density and Distribution

Due to the rich soil, waterways, and milder climate of western Russia, about 85% of the population live there. The major industrial city and capital, Moscow, is located in western Russia. Since 1990, population growth in Russia's industrial cities has leveled off or decreased.

East of the Urals, Siberia covers two-thirds of the country, but only 15% of the people live there. The frozen tundra, forests, and mountains make most of Siberia unsuitable for farming.

In the early Soviet era, many Russians moved to non-Russian republics. Since 1991, more ethnic Russians have returned to their homeland than have left.

Language

While more than 100 languages are spoken in Russia, Russian is the official language. In western Russia, Turks speak Altaic languages and Russian. In eastern Russia, the Sakha people speak Turkic languages and Russian.

DID YOU KNOW?

Maslenitsa was a traditional Russian folk celebration of spring absorbed by Eastern Orthodox Christianity. When the Soviet Union fell, the flood of Western Christian missionaries prompted Russian lawmakers in 1997 to only allow Russian Eastern Orthodoxy, Islam, Judaism, and Buddhism complete liberty as traditional religions of Russia.

Healthcare

Healthcare is declining in Russia. Life expectancy in Russia is 66. It is 78 in the United States. Russian infant mortality is 12 per 1,000 births, twice the rate of the United States.

Whereas the Russian death rate has exceeded the birth rate since 1992, Russia has negative population growth.

DID YOU KNOW?

The number of people with HIV or AIDS continues to grow at alarming rates throughout Russia. The majority of cases are contracted through intravenous drug use, another serious health issue in Russia.

Nationalism in Chechnya

Chechnya in southwest Asia was occupied by the Turks, then the Russians. Most Chechens are Sunni Muslims with their own language and culture.

Russia wants to keep Chechnya a part of the Russian Federation because gas pipelines vital to the Russian economy run through Chechnya. Also, Chechen independence would embolden other ethnic groups and republics in the Russian Federation to secede.

Russian occupation forces can be seen on the streets of the bombed-out capital of Grozny and throughout Chechnya. Chechen rebels are still fighting violently against Russia. Chechen rebels use the methods of terrorism in seeking independence from foreign occupation and domination.

History

Ivan IV, "the Terrible," became Russia's first supreme ruler, or **czar**, in 1547. He crushed all opposition to his rule and expanded into non-Slav territories. The Romanov dynasty came to power in 1613. By 1650, an enslaved workforce of peasant farmers, or serfs, bound to the land fell under the control of the powerful Russian land-owning nobles.

Romanov Czars and Empire

Peter I, "the Great," modernized Russia. He created a strong military, enlarged Russian territory, and traded with Western Europe. He acquired warm water seaports along the Baltic Sea from Sweden. He built a capital, St. Petersburg, as his "window on the West" with access to the Baltic Sea.

Catherine the Great acquired a warm water port on the Black Sea. As the Russian nobility adopted genteel European ways, the **serfs** followed folk traditions and suffered from poverty and hunger.

Czar Alexander III connected Moscow to Vladivostok on the Pacific Ocean by constructing the 6,000-mile Trans-Siberian Railroad in 1891. The world's longest railroad, the Trans-Siberian opened Russia's interior to settlement.

Russian Revolution

Czar Alexander II's 1861 abolition of serfdom without education reform prompted many serfs to move to cities. In cities, serfs faced poor wages and factory conditions. The government forced people to speak Russian and follow Eastern Orthodox Christianity. Those who refused, like Jews, were blamed for Russia's problems.

Socialist workers and thinkers called for economic equality. Karl Marx, a German Jew writing about the labor struggle in British cities, advocated a working class revolution against the wealthy followed by public ownership of land, equal sharing of wealth, and a classless society.

Rising discontent in the early 1900s and the hardships of World War I forced Russian workers and soldiers into the streets in 1905 and 1917, demanding "bread and freedom." Czar Nicolas II abdicated the throne and his family was murdered by the Russian workers' revolution.

Soviet Era

Lenin and the Russian **Bolsheviks** overthrew a weak representative government to create a society led by workers and its elite Communist Party. Promising "Peace, Land, and Bread!" the Bolsheviks surrendered territory to the Germans and withdrew from World War I. The Bolsheviks took over industry and food distribution and established an eight-hour workday.

A civil war erupted between the Bolshevik Red Army and the anti-Bolshevik, or Menshevik, White Army. The Bolsheviks won the civil war by capturing the heart of Russia and established the Union of Soviet Socialist Republics (USSR), or the Soviet Union, in 1922. Ukraine, Belarus, and 13 other constituent unions of Caucasus and Central Asia became a part of the USSR.

When Lenin died in 1924, the rude but cunning administrator Stalin seized leadership of the Communist Party despite Lenin's warnings. Stalin took over farms and factories and made Russia an industrial giant. Millions starved when Stalin **collectivized** agriculture. Stalin also murdered or imprisoned millions of opponents in brutal labor camps.

Superpower

The Soviet Union achieved superpower status after losing 20 million men and women in World War II, a war the Russians called *The Great Patriotic War*. The USSR moved to occupy Eastern Europe and North Korea as buffer zones at the end of the war. Most of these countries became Soviet-controlled **satellites**.

For the next forty years, the United States and Soviet Union used the ideologies of capitalism and communism to compete for world dominance. The **Cold War**, fueled by propaganda, conventional and **nuclear** threats of war, and aid to developing proxy countries, involved the two major world superpowers.

Fall of the Soviet Union

A weak economy and the income gap between those with Communist Party privileges and those existing on worker wages led to the breakup of the Soviet Union.

In 1985, Gorbachev instituted **perestroika**, or economic restructuring, and **glasnost**, or political openness, in Russia. Many of the Soviet satellites threw off communist rule by 1989. All the Soviet republics declared independence over the next two years.

Twelve of the fifteen Russian republics joined the **Commonwealth of Independent States (CIS)**. The Baltic States did not join the Commonwealth.

DID YOU KNOW?

In 1991, a hard-line communist coup to overthrow Gorbachev failed and Boris Yeltsin became the first democratically elected president of the Russian republic.

New Russia

After 1991, outdated factories were closed. Agriculture was restructured. Russia moved from a **command economy** to a market economy and inflation went from 1,500% in 1992 to below 20% in 1997. Tatarstan, Dagestan, and other ethnic territories demanded more self-rule.

Yeltsin's successor, Putin, has stabilized the economy since 1999 by instituting banking, labor, and private property reforms. Putin involved Russia in the North Atlantic Treaty Organization (NATO). Since 2004, Putin has expanded his executive power and curtailed democratic freedom in Russia.

Practice

Fill in the blank with the correct word from the information in the preceding paragraphs.

1. In the 600s, the ____________ settled the North European Plain.

2. ____________ is the largest ethnic group in Russia.

3. Peasant farmers known as ____________ were bound to the land of Russian nobles.

4. Rich soil, waterways, and the milder climate of ____________ draw about 85% of the population to live there.

5. Russia has the longest ____________ in the world.

6. An eight-hour workday was established by the ____________ after World War I.

7. Politically, Russia is part of the ____________ of Independent States today.

8. Since 1985, political openness or, ____________, has expanded in Russia.

9. Economic restructuring is called ____________ in Russia.

10. The religion of most people in southwest Asia's oil rich Chechnya is ____________.

Russia and the United States

Russian culture has influenced American arts, sports, religion, and space exploration. There are about 3 million Russian Americans.

Cold War

From the close of World War II in 1945 until the collapse of the Soviet Union in 1991, there was great tension between Russia and the United States. Allies during the war, the two superpowers disagreed about how borders should be restructured after the war. The two powers have never directly gone to war, but they have struggled for geopolitical, technological, and military superiority.

DID YOU KNOW?

The Korean War was the first military conflict of the Cold War. Committed to stopping the spread of communism, the United States supported South Korea, while Russia gave its support to the North.

The two nations became involved in a space race. The Soviets launched the first satellite into orbit around Earth in 1957, and the United States responded by realizing President John F. Kennedy's promise of beating the Russians by landing on the moon first in 1969. They also competed to develop the first nuclear weapons. The Cuban Missile Crisis of 1962 was the most direct threat of nuclear war, with Soviet atomic warheads stationed about 1,000 miles off American soil.

DID YOU KNOW?

The term *cold war* was first coined by English author George Orwell in the book *You and the Atomic Bomb*, published in 1945.

Agriculture

Soviet-era state-controlled farms, or **kolkhozes**, paid farmers as salaried employees and were owned by collectives. Other farms, called **sovkhozes**, were state-owned and also paid workers wages. Production and prices were still controlled by the government in agriculture and industry. The system did not motivate workers.

Although Yeltsin opened farms to market reform starting in 1991, most farmers could not afford to buy land. Fearing that wealthy Russians or foreign investors would buy the land and use it for nonagricultural purposes, farmers have been reluctant to stray from the familiar and stable kolkhoze or sovkhoze systems.

Industry

Industry and services are expanding in Russia. Russia is one of the world's largest producers of crude oil. Russia's oil supply provides vital energy at a reasonable cost. Energy and minerals provide Russia with income from exports.

DID YOU KNOW?

Russian forests provide 20% of the world's soft wood.

Supertrawlers, or floating fish factories, process catches from the Atlantic and Pacific oceans.

Transportation

Because of its immense size and climate extremes, Russia needs railroads and waterways for transportation. Major cities are located where the railroad crosses large rivers. Thousands of miles of navigable inland waterways connect seaports with cities.

Pipelines transporting petroleum to Russian cities and Western Europe crisscross Russia. The republics of southwest Russia, Chechnya, and Dagestan are fighting for independence and control of their oil reserves and pipelines.

People and Environment

Russia faces massive environmental challenges, including the repair of serious Soviet-era damages and making the best use of abundant natural resources for economic growth. Water, air, and soil have been harmed.

Russia must balance demand for oil and timber without causing further damage to the environment.

Nuclear Wastes

Russia stockpiled and then set off over 600 nuclear warheads between 1949 and 1987. **Radioactive** by-products of nuclear power will continue to be dangerous to people and the environment for thousands of years. Most of this **nuclear waste** was stored, but some of it was dumped into the Baltic, Barents, and Bering seas.

Chernobyl

In 1986, a fire at the Chernobyl nuclear reactor sent 400 times more radioactive material into space than was produced by the atomic bomb dropped on Hiroshima.

Radiation spread across tens of thousands of square miles. Thousands died and millions suffer from severe radioactive poisoning because the Soviet government was slow to evacuate people. Ultimately 35,000 people were displaced from their homes.

Russia still operates 29 nuclear reactors, despite international pressure for improved nuclear safety standards. Experts think the Russian nuclear reactors are still unsafe, yet Russia plans to expand the use of nuclear energy for electrical power. In 2000, the reactor at Chernobyl was shut down. In 2006, work began on a new protective structure for it.

Water Quality

Fertilizer runoff, sewage, and radioactive material have polluted the waters of Russia's lakes and rivers. The Moskva and Volga Rivers pose serious dangers to health. The Caspian Sea is threatened by industrial pollution. Since 1957, Lake Baikal's 1,500 species of native plants and animals have been harmed by the dumping of industrial waste from factories.

Soil and Air Quality

Russia's soil has been poisoned by decades of airborne pollution and the dumping of toxic waste. Storage containers crack and waste leaks into the soil. Petroleum pipelines break, ruining the land and water. Fertilizers and pesticides pollute farmland and water.

Russia has widespread urban air pollution. Vehicle, coal, and industrial emissions pollute the air. Burning coal releases soot, sulfur, and carbon dioxide into the air, forming acid rain. Acid rain and chemical pollution have severely reduced Russian forests.

Managing Resources

Russia is striving to repair damage to the environment and manage natural resources without causing additional harm.

The World Bank's Sustainable Forestry Pilot Project is showing Russia how to use land wisely, plant new trees, and invest in the environment and economy. Local economies depend on forest conservation.

Environmental groups have demanded that mining companies meet strict environmental standards. People have protested mining operations in Kamchatka, eastern Russia. The local fishing industry opposed the mines due to possible effects on the area's salmon spawning grounds, and local residents opposed the mine due to its proximity to a protected wildlife area. Frequent protests have reduced pollution in Lake Baikal, which holds 20% of Earth's freshwater. Pollution levels in the lake are now lower than in many lakes in Europe.

Future Challenges

Russia faces many challenges as the country's economy grows and demand for natural resources impacts the environment. Russian supertrawlers have depleted world fish stocks. They scoop up 400 tons of fish a day and discard the fish and marine mammals they do not want. Big ships threaten traditional or indigenous fishing cultures and fish stocks.

Oil and natural gas company pipelines threaten wildlife environments and indigenous ways of life in Russia like reindeer hunting.

Billions of dollars have been spent on an oil pipeline from eastern Siberia to the Pacific Ocean. Fearing oil spills, President Putin ordered the pipeline away from a wilderness area near Lake Baikal.

Global warming affects Russia. The world's largest peat bog is thawing and releasing billions of metric tons of methane, a greenhouse gas, into the atmosphere. Where the sub-Arctic was once permafrost, it is now shallow lakes.

Practice

Insert T or F to indicate whether the statement is true or false, based on what you've read.

11. _____ Russian supertrawlers limit the amount of fish they scoop up each day to under one ton.

12. _____ Russia is a major world supplier of soft wood.

13. _____ Russia still uses nuclear energy for electricity.

14. _____ The shallow lakes that are beginning to replace some of the permafrost in the sub-Arctic are a result of global warming.

15. _____ The Moskva River is polluted with fertilizer runoff, sewage, and radioactive material.

16. _____ Vehicle, coal, and industrial emissions are not major contributors to Russia's air pollution problem.

17. _____ Environmental groups are concerned about mines, pipelines, and toxic waste dumps in Russia.

18. _____ There was more radioactive material produced by the 1986 fire at Chernobyl than by the atomic bomb dropped on Hiroshima.

Answers

1. Slavs
2. Ethnic Russian
3. Serfs
4. Western Russia
5. Railroad
6. Bolsheviks or Communists
7. Commonwealth
8. Glasnost
9. Perestroika
10. Sunni Muslim
11. F
12. T
13. T
14. T
15. T
16. F
17. T
18. T

LESSON

13 EUROPE—PHYSICAL GEOGRAPHY

LESSON SUMMARY

Physical forces shaped the landforms, water systems, and natural resources of Europe. Latitude, mountain barriers, ocean currents, and proximity to large water bodies affect Europe's climate and vegetation. Such physical features still shape the lives of Europeans. In general, Europeans have prospered using the continent's natural resources.

The development of civilization and industry in general has always shown itself so active in the destruction of forests that everything that has been done for their conservation and production is completely insignificant in comparison.

—Karl Marx

Physical Geography of Europe

Europe is conventionally considered one of the seven continents, but is sometimes considered a subcontinent or large peninsula of northwestern Eurasia.

Europe is relatively contained, separated from the rest of Eurasia—with Asia to the east—by the Ural Mountains, the Caspian Sea, the Caucasus Mountains, and the Black Sea.

Landforms

Europe's landscape was created over time by the physical processes of wind, water, and ice. These processes have shaped the lives and settlement patterns of European people.

Mountains and Plains

The European landscape consists of plains with mountains on its northern and southern edges.

Mountains of northwest Europe were rounded by millions of years of erosion and **glaciation**. Glaciation is the result of spreading glaciers wearing down the earth.

The central uplands extend from the Iberian Peninsula, which encompasses Spain, Portugal, and Andorra, to Eastern Europe. The central uplands have low, rounded mountains and high plateaus with scattered forests.

The mountains of southern Europe are geologically young, high, jagged mountains. Formed by glaciation and folding, the Alps run from France to the Balkan Peninsula. The Alps are the source of some of Europe's major rivers, such as the Rhine and the Po.

DID YOU KNOW?

Mount Blanc is the highest point in the Alps at 15,771 feet.

The Alps form a barrier separating the warm, dry climate of the Mediterranean from the cooler northern climates. The Pyrenees mountain range on the Iberian Peninsula rises 11,000 feet high. The towering Carpathians run through Eastern Europe from Slovakia to Romania.

In Europe, broad plains curve around the highlands. Scoured by Ice Age glaciers, the North European Plain stretches from southeastern England and western France east to Poland, Ukraine, and Russia. The North European Plain is a highly productive agricultural area with mild climate, fertile soil, and access to rivers. The southern edge has particularly fertile soil enriched by organic glacial runoff called **loess**.

DID YOU KNOW?

Coal, iron ore, and other mineral deposits near the North European Plain provided the basis of Western Europe's industrial development in the 1800s.

Many of Europe's largest cities, like Berlin and Paris, were built on the North European Plain.

The Great Hungarian Plain extends through Croatia, Serbia, and Romania. On the lowlands along the Danube River, farmers cultivate grains, fruits, and vegetables.

Peninsulas, and Islands

Europe has a long, unusual coastline that touches many bodies of water, including the Atlantic Ocean and the Baltic, North, Mediterranean, and Black seas. Most of Europe is less than 300 miles from a seacoast.

DID YOU KNOW?

About 25% of the Netherlands lies below sea level. Since the Middle Ages, the Dutch have built large earth and stone embankments called **dikes** to hold back the water.

The Dutch have reclaimed land from the sea. They call this land **polders**. The Dutch drain the land and keep it dry using windmills, pumps, and other power sources. Although hundreds of thousands of acres of polders have been rescued for farming and settlement, stormy seas still breach the dikes, causing flooding.

Northern Peninsulas

Glaciation carved narrow, steep-sided **fjords** along the coasts of Europe's northern peninsulas, particularly along Scandanavia, and created thousands of sparkling lakes in Norway, Sweden, and Finland.

Northern Sweden and most of Norway are mountainous. Warm ocean currents create a marine west coast climate for coastal areas of the northern peninsulas.

The Jutland Peninsula forms mainland Denmark and extends into the North Sea. Glaciers carved fjords on the eastern coastline and deposited sand and gravel on the flat west side. The Jutland interior has low hills and flat plains.

Southern Peninsulas

Southern European peninsulas include the Iberian, Italian, and Balkan Peninsulas. Mediterranean climate characterizes the Iberian and Italian Peninsulas.

The Iberian Peninsula extends from southwest Europe, separating the Atlantic Ocean from the Mediterranean Sea. Most of Iberia is a plateau with coastal plains.

DID YOU KNOW?

The Strait of Gibraltar is a stretch of water less than 10 miles long separating Spain's southern tip from Morocco in North Africa. The Atlantic Ocean is on one side and the Mediterranean Sea is on the other.

The Pyrenees mountains in the Iberian north isolated people from the rest of Europe and oriented them toward the sea.

The Italian Peninsula extends like a boot into the Mediterranean Sea. The long coastline has high, rocky cliffs in some places and sandy beaches in others. The young Apennine Mountains form the Italian peninsula's spine. Plains make up a third of the Italian Peninsula. Lombardy along the Po River in the north is the Italian Peninsula's most fertile plain.

Southeastern Europe's Balkan Peninsula has the Adriatic and Ionian seas on the west and the Aegean and Black seas in the east. Craggy mountains and valleys stretch south from the Danube River, so people of the area usually travel by river or sea.

Europe's Islands

Iceland lies south of the Arctic Circle along the Mid-Atlantic Ridge. Iceland has many volcanoes, glaciers hot springs, and geysers. These sources of geothermal power are often harnessed for home and industrial heating.

DID YOU KNOW?

Nearly a quarter of Iceland's population died after the Laki volcano erupted in 1783 and set off a famine.

Iceland has tundra and marine west coast climates. Iceland's lowland coasts rise to an inland plateau.

The British Isles, northwest of mainland Europe, consists of two major islands, Great Britain and Ireland, as well as thousands of small, beautiful islands. The rugged rocky cliffs of the coastlines drop to deep bays in the British Isles.

Most of north and western Great Britain are made up of mountains, plateaus, and valleys. Low, rolling hills dominate the south. Ireland, the Emerald Isle, is a lush, green land with cool temperatures and abundant rainfall.

There are islands south of mainland Europe in the Mediterranean Sea. Rugged mountains form the islands of Sicily, Sardinia, Corsica, Crete, and Cyprus.

DID YOU KNOW?

Europe's tallest active volcano, Mount Etna, rises nearly 11,000 feet above Sicily.

Smaller island groups in the region include Malta's five islands, Spain's Balearic Island, and Greece's 2,000 islands in the Aegean Sea.

Water Systems

Europeans depend on rivers and canals for transportation, trade, and leisure activities.

Rivers and Lakes

Many of Europe's rivers run from mountains or highlands to the coasts. Canals link Europe's interior to its many navigable rivers. In Europe's heartland, long rivers link cities and run to the sea.

The most important river in Western Europe, the Rhine, flows from the Swiss Alps through France and Germany to the Netherlands. The Rhine connects many industrial cities to Rotterdam on the North Sea. The Main River, a tributary of the Rhine, connects to the Danube River by canal and links the North Sea with the Black Sea.

The Danube River is Eastern Europe's major waterway. The Danube flows from Germany's Black Forest through Hungary and Romania to the Black Sea.

DID YOU KNOW?

Europe's dominant Rhine and Danube Rivers carry hundreds of millions of tons of cargo every year.

France's Seine, Rhone, and Loire rivers are important for transportation and urban development. The Po River in Italy is important for industrial development. Europe's rivers also provide irrigation and electrical power.

European rivers differ region-to-region. Scandinavian rivers are short and rarely connect cities. Iberian rivers are too narrow and shallow for large ships. However, England's Thames River permits oceangoing ships to reach the port of London.

Natural Resources

Natural resources influence economic activity in Europe. Europe's abundant coal and iron ore fueled development of modern industry.

The North Sea contains major petroleum and natural gas reserves that meet Europe's energy needs. The United Kingdom, Germany, Ukraine, and Poland also have significant coal deposits.

Lacking large oil and gas reserves, France has invested heavily in nuclear power. Vegetable matter from swamps called peat is also dried and burned as an energy source. Sweden, France, and the Ukraine have iron ore. Minerals, including bauxite, zinc, and manganese, are also found in Europe.

Practice

Fill in the blank with the correct word from the information in the preceding paragraphs.

1. The ___________ Sea contains major petroleum and natural gas reserves that meet Europe's energy needs.

2. Europe's dominant ___________ and ___________ rivers carry hundreds of millions of tons of cargo every year.

3. Natural resources such as __________ and __________ __________ fueled development of modern industry.

4. The British Isles are made up of two major islands, __________ and __________.

5. Spain and Morocco are separated by a narrow stretch of water called the __________.

6. Mountains of northwest Europe were rounded by millions of years of __________ and __________.

7. Three nations, __________, __________, and __________, are located on the Iberian Peninsula.

8. Glaciation carved __________ on Denmark's eastern coastline.

9. The __________ form a barrier separating the warm, dry climate of the Mediterranean from the cooler northern climates.

10. Croatia, Serbia, and Romania make up the Great __________ Plain.

Influences on Climate

Latitude, large mountain barriers, wind patterns, and distance from large bodies of water influence Europe's climate.

Europe's climate varies from the cold subarctic barren tundra of Iceland, Norway, Sweden, and Finland to the warm, olive tree-covered Mediterranean areas of Greece, Italy, and Spain.

Western and southern Europe lie near large bodies of water. These areas receive warm maritime winds brought by the North Atlantic current. These winds provide a milder climate than other places at the same latitude receive. Therefore, Paris is often warmer than Boston in winter.

> **TIP**
>
> Eastern Europe may be colder due to distance from the warm winds of the Atlantic Ocean, but it also gets less precipitation.

Location affects vegetation patterns in Europe. Climate regions affect whether an area will have forests, grasslands, tundra plants, or small shrubs.

Climate Regions

Europe's climate regions range from the cold northern climates of the Alps to the dry steppe and Mediterranean climates found in southern parts of the continent.

High Latitude Regions

The far north's high latitude climates have bitterly cold winters and short summers, causing permafrost deep below the tundra's surface. Tundra supports little vegetation except for mosses, shrubs, and wildflowers that bloom during the brief summer. The subarctic climate, however, supports a vast conifer forest that broadens in the east where Europe and northern Asia meet. Fir, pine, and spruce trees fill this rugged northern landscape.

Western Norway, southern Iceland, and Sweden have a warmer midlatitude-type climate. This is because the North Atlantic and Norwegian currents provide a marine west coast climate with cool summers, mild winters, and nearly 90 inches of annual precipitation. These tropical water currents keep Norway's jagged coastlines, or fjords, from freezing.

Parts of eastern Norway, southern Sweden, and Finland that are sheltered by mountains have a humid continental climate. In this climate, warm summers, cold winters, and less than 30 inches of annual precipitation prevail.

Midlatitude Regions

Most of Western Europe has a marine west coast climate with mild winters, cool summers, and abundant rainfall. The Gulf Stream and North Atlantic currents bring warm waters from the Gulf of Mexico and the equator. Prevailing westerly winds blowing over these currents brings warm, moist air to Western Europe.

Natural vegetation of the midlatitudes includes deciduous and coniferous trees. Deciduous trees that lose their leaves, like ash, maple, and oak, thrive in the area's marine west coast climate. Coniferous trees like fir, pine, and spruce are found in cooler Alpine mountain areas up to the **timberline**. After that point, trees cannot grow.

Southern Europe has a Mediterranean climate with hot, dry summers and mild, rainy winters. The humid subtropical climate extends from northern Italy to the central Balkan Peninsula.

Hot, dry winds from North Africa sometimes bring high temperatures to southern Europe. Such heat and lack of precipitation encourage the growth of drought-resistant vegetation like shrubs and small trees.

Most of Eastern Europe has a humid continental climate with cold, snowy winters and hot summers. This region is farther from the influence of the Atlantic Ocean's warm waters. The region's humid continental climate supports vegetation that is a mixture of deciduous and coniferous forests.

The Alps have a highland climate with colder temperatures and more precipitation than nearby lowland areas. Dry winter winds called **foehns** blow down mountains into valleys and plains. These foehns cause **avalanches**—destructive masses of ice, snow, and rock that slide with tremendous force down mountainsides.

Dry Regions

Parts of southeastern and southwestern Europe have a dry, steppe climate. Southeastern Europe's steppe extends through Serbia, Hungary, Montenegro, Romania, Ukraine, and Central Asia. The steppe has hot summers and extremely cold winters. Precipitation varies. Rainfall is scarce far from the Atlantic Ocean's winds in the eastern steppe. Extreme temperatures, droughts, poor and easily eroded soils, and high winds make farming the steppe very difficult.

The interior Iberian plateau, the Meseta, has a dry steppe climate. The Meseta extends over 81,000 square miles. Madrid is at its center.

Practice

Insert T or F to indicate whether the statement is true or false, based on what you've read.

11. ____ The European steppe is easily farmed.

12. _____ Coniferous trees are common in the warm Mediterranean climate of Southern Europe.

13. _____ Permafrost is found in the midlatitudes.

14. _____ Western and Southern Europe have milder climates due to winds brought by warm ocean currents.

15. _____ Foehns cause avalanches in the Alps.

16. _____ Fir, pine, and spruce trees are found at seal level near European coastlines.

17. _____ Temperatures in Southern Europe are affected by winds from North Africa.

18. _____ The Mediterranean climate supports a vast conifer forest that broadens in the east where Europe and northern Asia meet.

19. _____ Most of Eastern Europe has a humid continental climate with cold, snowy winters and hot summers.

20. _____ Warm North Atlantic and Norwegian currents keep Norway's fjords from freezing.

Answers

1. North
2. Rhine and Danube
3. Coal and iron ore
4. Great Britain and Ireland
5. Strait of Gibraltar
6. Erosion and glaciation
7. Spain, Portugal, and Andorra
8. Natural gas
9. Alps
10. Hungarian
11. F
12. F
13. F
14. T
15. T
16. F
17. T
18. F
19. T
20. T

EUROPE—HUMAN GEOGRAPHY

LESSON SUMMARY

The people of Europe belong to many different cultural groups and speak a variety of languages. Throughout its long history, Europe has overcome the challenges associated with blended cultures.

Europe is so well gardened that it resembles a work of art, a scientific theory, a neat metaphysical system. Man has re-created Europe in his own image.

—Aldous Huxley

Countries

Albania
Andorra
Austria
Belarus
Belgium
Bosnia and Herzegovina
Bulgaria
Croatia
Cyprus
Czech Republic
Denmark
Estonia
Finland
France
Germany
Greece
Hungary
Iceland
Ireland
Italy
Latvia
Liechtenstein
Lithuania
Luxembourg
Macedonia
Malta
Moldova
Monaco
Montenegro
Netherlands
Norway
Poland
Portugal
Romania
San Marino
Serbia
Slovakia
Slovenia
Spain
Sweden
Switzerland
Ukraine
United Kingdom
Vatican City

Northern Europe

Northern Europe is made up of the United Kingdom, Ireland, and Scandinavia.

Population Patterns

The United Kingdom, which includes England, Scotland, Wales, and Northern Ireland, is ethnically diverse and densely populated. Ireland, Iceland, Norway, Sweden, Finland, and Denmark have smaller populations dominated by one or two ethnic groups.

People

The British Isles are among the most diverse and densely populated areas in Northern Europe. Celts, or people who spoke Celtic languages, migrated from Eastern Europe at least 3,000 years ago, as did Indo-Europeans, who came from Central and Eastern Europe and South and Central Asia. The Romans, Normans, and others followed.

Since 1900, large numbers of immigrants from the West Indies (the Caribbean) and South Asia have arrived on the islands of Northern Europe. Many European refugees fleeing for their safety settled in Great Britain at the end of World War II.

Scandinavians share a Viking and Germanic heritage and similar ways of life. Swedes, Norwegians, and Danes speak different, but related, languages.

Density and Distribution

The most densely populated country in Northern Europe is the United Kingdom, with roughly 660 people per square mile. Denmark and Ireland also have high population densities. All three countries have temperate climates and fertile soil that support large populations.

Northern Europe's important metropolitan areas are also its economic centers. Some examples are: London, England; Paris, France; Stockholm, Sweden; and Copenhagen, Denmark.

Northern Europe Rises

The Romans added Britain to their empire in A.D. 43. They built roads, cities, and towns and brought Christianity. The Roman Empire declined starting in the A.D. 300s, and the Germanic Angles, Saxons, and Jutes invaded Britain when the Romans left in the early Middle Ages from A.D. 500 to 1500.

Britain fell to Norman invaders from France in 1066, and the Norman king established a system of monarchs, or lords, giving lands to nobles for loyalty pledges. Towns and villages formed to serve kingdoms.

The religious Reformation of the 1500s, which began as an attempt to reform the Roman Catholic Church, established Protestantism as a branch of Christianity. The reform movement was led by a German monk, Martin Luther.

Powerful Vikings from Scandanavia raided the European coastlines from the eighth to eleventh centuries. The Viking kingdoms provided the foundations for Sweden, Norway, and Denmark.

Language and Religion

Northern Europe's languages, like English and Swedish, are mostly Indo-European. Local dialects exist. Most North European countries are Protestant, but many minority religions exist.

Education and Healthcare

Northern Europe has some of the world's most educated populations. School is mandatory for at least 10 years and literacy rates are nearly 100%. Sweden, a social democracy, offers complete healthcare and social services to its citizens.

Western Europe

The Western European countries form a cultural subregion influenced by its location, history, and revolutionary role in world events. It is a crossroads of cultures and a mix of the old and the new.

Population Patterns

Western Europe is one of the smallest but most populous regions of the world. It has long been and still is a place of migrations.

People

The people of France, Belgium, the Netherlands, and Switzerland are blends of two or more ethnic groups that mixed over the centuries. Germany and Austria were in ancient times ethnically homogeneous, but now many ethnic groups live in each country.

France has recently attracted many Muslim immigrants from Algeria and Morocco.

Density and Distribution

Most of Western Europe is densely populated, particularly in urban areas. Germany has the largest population in Europe at 82.5 million people, while the small urban countries of the Netherlands and Belgium have the highest population density.

Migration in Europe is mostly to urban or suburban areas. Since World War II, immigrant **guest workers** began flocking to European cities.

History and Government

The history and governments of Western Europe have been shaped by early civilizations, revolutionary politics, world wars, and migrations.

Early Peoples

Celts and other ancient peoples populated Western Europe. Romans controlled their lands for centuries until Germanic conquerors overpowered the Roman Empire.

The Basques have lived in the Pyrenees between France and Spain since more than 3,000 years ago, well before the Celts arrived. The seafaring Frians settled in Netherlands on the North Sea around 400 B.C.

Western Europe Rises

After the Roman Empire fell, the Franks adopted a Roman form of Christianity later known as Roman Catholicism. From A.D. 768 to 814, Charlemagne expanded Christianity and the manorial land distribution system known as **feudalism.**

Western Europe grew in power in the late **Middle Ages** (A.D. 1000–1500) and came into contact with other regions. Starting in A.D. 1000, European armies fought **Crusades** against Muslims to control Palestine, the birthplace of Judaism, Christianity, and Islam.

In the late 1400s, educated Europeans engaged in a popular renaissance of interest in classical Greek and Roman ideas. Such ancient ideas were spread across Europe by merchants and visitors from Italy.

Winds of Change

In the struggle for territorial and religious power, mostly Catholic France and Protestant England fought the Thirty Years War (1618–1648). France emerged as a major world power.

In the 1700s, the **Enlightenment** inspired democracy and social change. The British Empire was defeated by a revolution of its colonies in North America. The French monarchy was overthrown by the French Revolution. France restored monarchy by 1815, but kings would never again wield absolute power.

In the mid-1800s, the kingdom of Prussia unified Germany with a strong industrial base and expanding military. The Central and Allied powers diplomatically, industrially, and militarily aligned themselves in a struggle for world dominance that pushed Europe into World War I (1914–1918).

Due to new methods of chemical, artillery, machine gun, and aviation warfare, casualties in "the Great War" were enormous. Having lost, Germany was assigned blame for starting the war and was forced to pay reparations for damages. German aggression led to World War II, which lasted from 1939 to 1945. Millions of lives were lost, and the borders of many countries were redrawn. Germany, Finland, Poland, Czechoslovakia, and Romania all gave up land to an expanding Soviet Union.

DID YOU KNOW?

The United Nations was developed after World War II to prevent future worldwide conflicts. The World Bank and **World Trade Organization (WTO)** were also developed in response to World War II.

New Era

After World War II, Germany was divided into communist East Germany and democratic West Germany. The Berlin Wall prevented movement between the two sides. The divide remained until the fall of the Soviet Union in 1991.

In the 1950s, West Germany, the Netherlands, Luxembourg, France, Italy, and Belgium grew closer together politically and economically.

Arts

Western Europe has been at the forefront of literature, architecture, music, and visual arts. **Realism** became a prominent artistic movement in the mid-1800s as a response to the emotional style of **Romanticism.** Later, the Impressionists moved outdoors to capture immediate impressions of the natural world.

Language and Religion

The three official Indo-European languages of Switzerland—German, French, and Italian—reflect the influences of three unique cultures.

DID YOU KNOW?

The Basque language of Euskera is one of the few languages that is not related to any other.

Western Europe is predominantly Christian. A vast majority are nominal Roman Catholics who engage in traditional celebrations.

Education and Healthcare

Just like in Northern Europe, Western Europe has a tradition of compulsory education and comprehensive healthcare and social services. Germany's social welfare system is typical in providing citizens with unemployment benefits and other services. About 98% of Western Europe is literate.

Eastern Europe

Eastern Europe has emerged from centuries of power struggles with a rich culture and growing economic strength.

Population Patterns

Eastern Europe's population patterns have been shaped by diverse physical geography, migration, and political and ethnic struggles. After the fall of communism, this region experienced a dramatic increase in emigration.

People

Most Eastern Europeans are ethnic Slavs. Slavs descended from Indo-Europeans who migrated from Asia. Eastern Slavs are from Russia, Ukraine, and Belarus. Western Slavs include Poles, Czechs, and Slovaks. South Slavs are the Serbs, Croats, Slovenes, and Macedonians.

The central part of Eastern Europe is sometimes called the Balkans. It includes the former countries of Yugoslavia: Serbia, Montenegro, Bosnia and Herzegovina, Croatia, Slovenia, and Macedonia. Southern Slavic people here are Eastern Orthodox Serbs, Roman Catholic Croats, and Bosnian Muslims. Further north are the Czech Republic, Poland, Hungary, and Slovakia.

The eastern part of Eastern Europe includes Slavic Russians and Ukrainians. The Baltic Sea countries of Estonia, Latvia, and Lithuania also border on Russia.

Density and Distribution

Ukraine and Poland's fertile soil and ample water support large populations. Ukraine has the subregion's highest population density of 202 people per square mile.

Eastern Europe has experienced large-scale internal migration and emigration. Throughout the 1900s, people migrated from rural areas to cities in pursuit of industrial jobs. After World War II, many Polish people emigrated or left Poland to escape Soviet control.

Most of the population of Eastern Europe lives in large towns or cities. Since the end of Soviet occupation in the late 1980s, many Eastern European cities like Budapest, the capital of Hungary, have been reborn as business and cultural centers.

History and Government

The history and governments of Eastern Europe have been shaped by political, economic, and ethnic struggles, most recently by the dissolution of the Soviet Union in the mid-1990s.

Early Peoples and Empires

Slavs migrated from Asia to Ukraine and Poland thousands of years ago and lived among migrating Celts and Germans. Starting in the 400s, Germanic and Slavic groups moved south and west. Slavic Czechs settled Bohemia in the 500s and Central Europe's Great Moravia by the 700s.

Slavs in the mountainous Balkans settled the independent states of Croatia, Serbia, and Slovenia. In the 1400s, the Ottoman Empire named the area the Balkans, or "mountains."

DID YOU KNOW?

The Muslim Ottoman Turks ruled the Balkans for 500 years.

The eastern portion of the Roman Empire that remained for 1,000 years after the fall of Rome became known as the Byzantine Empire. Byzantine missionaries spread Eastern Orthodox Christianity across Eastern Europe.

Conflict, Union, and Division

Balkan Slavs overthrew the Ottoman Empire in the early 1900s. Yugoslavia, or "Land of the Southern Slavs," has not remained unified. The area **balkanized,** or broke up into smaller subdivisions with different ethnic and religious groups becoming hostile to one another.

In the 1990s, ethnic hatred sparked violence in Bosnia and Herzegovina and in Kosovo. Serb leaders killed or expelled rival ethnic groups in the areas in a process called **ethnic cleansing**. International peacekeepers have helped many refugees return home.

New Era

From the 1950s to the 1980s, there were revolts in Eastern Europe against communist rule. In 1989, public demonstrations brought down the communist governments of Eastern Europe.

Throughout the 1990s, free elections brought democratic leaders and market economics to power in Eastern Europe. Many Eastern European countries recently joined the **European Union (EU)**.

Language and Religion

Most Eastern Europeans speak Indo-European languages. Common Slavic languages include Polish and Czech. Baltic languages include Lithuanian and Latvian.

Roman Catholicism, Eastern Orthodoxy, and Islam are all common religions in Eastern Europe.

Education and Healthcare

Because school is mandatory and free of charge in Eastern Europe, literacy rates are high. Former Soviet satellites faced funding challenges in their transitions to democratic governments. Although the healthcare system was challenged in the 1990s, most Eastern Europeans have access to healthcare.

Southern Europe

The centuries-old cultures of Southern Europe, including Italy, Spain, Andorra, Greece, and Portugal, created modern industrial nations with distinct lifestyles.

Population Patterns

After decreasing steadily over the past 20 years, the populations of several Southern European cities, including Madrid, Barcelona, Rome, and Milan, are again on the rise.

People

The Mediterranean countries of Southern Europe have long been home to seafaring peoples and thriving civilizations. The Minoan (island) and Mycenaean (mainland) Greeks trace their language and culture back more than 3,500 years. The tribal Etruscans and Romans established ancient civilizations in Italy.

The prehistoric Iberians, who named their peninsula, settled Spain. Speakers of the Romance language of Catalan settled northeast Spain and the state of Andorra. At the western edge of Iberia, Portugal's coastline made seafaring accessible.

Density and Distribution

Southern Europe's long history of emigration includes millions of people leaving for America in the 1800s and 1900s. Most recently, immigration and internal migration to cities has surpassed external migration.

With approximately 505 people per square mile, Italy is the most densely populated country in Southern Europe.

DID YOU KNOW?

Located within Italy's capital of Rome, Vatican City is the world's smallest independent state and home to the Roman Catholic church.

History and Government

Shortly before its fall to Germanic tribes in the 400s, Christianity became the official religion of the Roman Empire. The eastern half of the empire survived as the Byzantine Empire. The capital of Byzantium was Constantinople, later called Istanbul.

Germanic tribes that sacked Rome and overthrew the Empire maintained Roman Catholicism. Italy remained fragmented for hundreds of years. The Muslim Moors invaded southern Spain in A.D. 711 and held it for 700 years.

The wealth and stability of Italian city-states led to the **Renaissance**, a rebirth of ancient Greek and Roman intellectual and artistic ideas, in the 1300s. People who had the freedom of leisure time pursued new ways of thinking. The vibrant new ideas and beautiful art of the wealthy Italian **city-states** attracted visitors. In the 1400s, Europe started exploring and colonizing the world. In the 1500s, Spain and Portugal achieved wealth from trade routes and colonies.

Winds of Change

In the 1800s and 1900s, Southern Europe experienced independence, nationalism, civil wars, dictatorships, and world wars, and has struggled through political and economic instability. Spain and Portugal lost most of their territories. Greece won independence from the Ottoman Turks. The Italian territories united in 1870.

Southern Europe was a major battleground in World Wars I and II. Greece was ruled by a military dictatorship from 1967 to 1974.

Many of the countries in Southern Europe now have democratically elected governments and membership in the EU.

Education and Healthcare

Due to strong government support for education, the literacy rate in Southern Europe is over 95%. Healthcare in Southern Europe depends on the country. Spain has a high number of doctors per person, while Greece does not have adequate healthcare. Government funds all healthcare programs and other social services in Southern Europe.

DID YOU KNOW?

The first European universities were located in medieval Italy.

Language and Religion

The Romance languages of Italian, Spanish, and Portuguese are Indo-European languages derived from Latin.

Most of the people in Italy, Spain, and Portugal are Roman Catholic. Most people in Greece belong to the Greek Orthodox Church.

Arts

Ancient Greeks and Romans developed columns, arches, and domes, basic elements of Western architecture. Italian Renaissance artists Leonardo da Vinci and Michelangelo, and Spanish modernist Pablo Picasso are part of the great Southern European artistic tradition.

Practice

Fill in the blank with the correct word from the information in the preceding paragraphs.

1. With roughly 660 people per square mile, ___________ is the most densely populated country in Northern Europe.

2. Southern European countries have a ___________ tradition.

3. Northern and Western Europe have a(n) ___________ population with declining birth rates and death rates.

4. Most Eastern Europeans are ethnic ___________.

5. The ___________ brought Christianity to Britain in AD 43.

6. Since the end of World War II, Western European countries have experienced a surge of ___________ immigrants.

7. In terms of religion, most North European countries are ___________.

8. Natives of three Scandinavian nations, ___________, ___________, and ___________, speak different but related languages.

9. The Muslim ___________ Turks ruled the Balkans for 500 years.

10. Greece, Italy, Portugal, and Spain are part of ___________ Europe.

Manufacturing

Large deposits of coal and iron ore sparked the development of heavy machinery and industrial equipment in Europe in the 1800s. Germany's Ruhr and Middle Rhine, France's Lorraine-Saar, Italy's Po basin, and Poland and the Czech Republic's Upper Silesia-Moravia are Europe's leading industrial centers.

Agriculture

Europe has fertile farmland. The percentage of agricultural workers in Europe varies by country from 2% in the United Kingdom to 48% in Albania.

Olives, citrus fruits, dates, and grapes grow in warm Mediterranean areas. In cooler northern plains regions, farmers grow wheat, rye, and other grains, and raise livestock.

Farming Techniques

Most Western European farmers own their own land of more than 30 acres. They use advanced technology to succeed in limited space. Mixed farming of several crops and livestock is common. Danish and other farmers form **farm cooperatives** to reduce costs, sell products, and increase profits.

After communism failed, farming changed in Eastern Europe. The lack of incentive and the outdated equipment of communism were replaced by democracy, private ownership of land, and increased food production yields and higher profits.

Agricultural Issues

Many Europeans are concerned about the safety of **genetically modified foods** and chemicals used to kill bugs and weeds. **Organic farmers** use natural products to increase crop yields. Western Europe is susceptible to livestock disease.

Agricultural **subsidies** are given to farmers to raise their income, help develop the agricultural industry, and protect agricultural prices. Opponents of agricultural subsidies say they cause crop overproduction and deflate prices.

Railways and Highways

Rail lines connect Europe's cities, airports, natural resources, and industrial centers. In 1981, France introduced high-speed rail lines that saved energy and were better for the environment. Germany, Italy, and Spain then constructed high-speed rail lines.

In 2000, Sweden opened a rail and road bridge connecting it to Western Europe. Paris, Brussels, and London are connected by a 31.4 mile-long tunnel underneath the English Channel. This **channel** tunnel, often called the chunnel, took 15,000 workers about seven years to construct, and transports around 15 million passengers per year.

Europe has a highly developed highway system connecting its major cities.

DID YOU KNOW?

Europe is second only to the United States in number of automobile owners.

Seaports and Waterways

Europe's long coastline gave it a **maritime** tradition. Today Europe handles half the world's international shipping. Rotterdam, Netherlands, is the world's largest and busiest port.

Europe's numerous navigable rivers and canals save transport costs. The Rhine River and its tributaries carry more freight than any other European river. Cities along the Danube River depend on it for trade. The Main-Danube Canal links inland ports between the North Sea and Black Sea.

People and Environment

Suffering the effects of damages caused by industry and development, Europeans are identifying environmental challenges, devising new ways of managing natural resources, and reversing the damages.

Flooding

The natural climate cycle and global warming have recently caused extremely heavy rains, flooding, and mudslides. In Western Europe, violent Atlantic Ocean and North Sea storms strike coastlines. After 1,800 people were killed by a flood in the Netherlands in 1953, engineers spent 30 years constructing a system of dams and dikes to seal off and protect its southwest coast.

Soil Erosion

Overfarming, removing too much vegetation, and overgrazing livestock have led to soil erosion in Europe. Long-term destruction of forests has caused soil erosion in the Mediterranean basin and the highly populated sandy coastal areas. **Reforestation** is a solution.

Deforestation

In Western Europe and the Mediterranean, trees were removed to build cities and farms. Northern Europe has healthy commercial forests.

DID YOU KNOW?

Over 80% of Europe was once forest. Two-thirds of it has been removed.

Many European countries are replanting trees or reforesting. Some countries are responsibly cutting trees. Since the 1800s, Sweden has maintained a strict system of cutting and replenishing trees because it takes 70 years to replace a full-grown spruce or fir in the southern part of Sweden and 140 years to replace one in the northern part.

Human Impact

Population growth and industrialization have impacted Europe's environment.

Europe's concentration of industry has devastated the environment in the heavily industrialized "Black Triangle" of eastern Germany, Poland, and the Czech Republic, where the air smells of sulfur from smokestacks and soot covers the ground.

Before 1989, communist Eastern European countries emphasized rapid industrial growth. There were virtually no pollution control laws. Western Europe has experienced environmental damage from the dumping of industrial wastes into the air and water.

Acid Rain

In the 1970s and 1980s, industrial smokestacks carried acid-producing chemicals into the air where they combined with moisture. Polluted clouds drifted from the European industrial belt and dropped acid rain on forests.

DID YOU KNOW?

Many Western European countries have shifted from coal to natural gas to reduce acid rain.

Many Eastern European countries still rely on coal and produce acid rain.

Snow carries pollution to the ground. In spring, meltwater carries acid into lakes and rivers, causing fish and aquatic life to die. Many lakes in Scandanavia have declining fish populations or no fish at all. Some rivers in Slovakia and the Czech Republic cannot support aquatic life.

Automobile exhaust adds acid-forming compounds to the atmosphere. Wet or dry acid pollution that is deposited on the ground harms the European environment, historic buildings, bridges, statues, and stained glass windows.

The EU sets strict emissions regulations for vehicles and industries. Smokestacks and vehicle exhausts must often be equipped with sulfur and nitrogen compound removal devices. Many people believe fossil fuels should be replaced by alternative energy sources like solar power.

Air and Water Pollution

Air pollution in the form of traffic exhaust and industrial fumes causes asthma, respiratory infections, and eye irritations for people living in European industrial areas. Eastern European factories built in the communist era belch sulfur, soot, and carbon dioxide into the air. Former communist countries are closing polluting factories, but they are putting more cars on the road that create traffic and pollution.

Growing populations along the coastline have been polluting the Mediterranean Sea by dumping industrial waste, sewage, garbage, and other pollutants in the water.

DID YOU KNOW?

Because the Mediterranean Sea opens to the Atlantic Ocean only through the narrow Strait of Gibraltar, it will take the polluted Mediterranean Sea about a century to renew itself.

Pollution harms marine and animal life and is dangerous to the health of humans. The Mediterranean is also overfished.

Europe's rivers and lakes also suffer from pollution. The Danube River is affected by algae growth that deprives the river of so much oxygen fish cannot survive. Industries in Western Europe deposit raw sewage into rivers like the Meuse and the Rhine.

Global Warming

The EU has supported the Kyoto Protocol, an amendment to the international treaty on climate change designed to reduce the amount of greenhouse gases emitted by each country.

Future Challenges

The EU requires environmental protection and thus takes legal action. Industrial countries are regulating pollution. Cities in Western Europe now put acid-resistant coatings on buildings and statues.

Danube River pollution is crossing national borders and threatening wildlife in its outlet, the Black Sea. Many European countries must cooperate to direct and finance water quality improvement.

Many industrial power plants are burning cleaner natural gas instead of coal. In 2005, Sweden introduced the first **biofuel**-powered passenger train.

DID YOU KNOW?

Biofuel is made of decomposing organic materials (biomass) that are much less dangerous to the environment than fossil fuels.

Eastern European countries are seeking financial aid from Western European countries to clean up and meet EU environmental standards. Western European and U.S. companies are providing technology and investment to help modernize Eastern Europe's industries.

Practice

Insert T or F to indicate whether the statement is true or false, based on what you've read.

11. _____ Spanish, French, and Italian are Indo-European Romance languages.

12. _____ The Danube carries more freight than any other European river.

13. _____ In warm Mediterranean regions, farmers grow wheat, rye, and other grains and raise livestock.

14. _____ Since its formation, the EU has imposed standards on trade, banking, and business law as well as environmental and human rights issues.

15. _____ Urban growth can be controlled by high-speed, high-frequency rail lines and automatic driverless trains that complement the existing bus system.

16. _____ The European Union encourages a single European currency, a central bank, and economic and foreign policy.

17. _____ The European Union does not require environmental protection and cleanup by members.

18. _____ Sweden's strict system of cutting and replenishing trees dates back to the 1800s.

19. _____ Many Western European countries have shifted from natural gas to coal in order to reduce acid rain.

20. _____ By forming cooperatives, farmers not only facilitate sales but also reduce costs and increase profits.

Answers

1. The United Kingdom
2. Seafaring
3. Aging
4. Slavs
5. Romans
6. Muslim
7. Protestant
8. Sweden, Norway, and Denmark
9. Ottoman
10. Southern
11. T
12. F
13. F
14. T
15. T
16. T
17. F
18. T
19. F
20. T

LESSON 15

AFRICA—PHYSICAL GEOGRAPHY

LESSON SUMMARY

Africa south of the Sahara is a region of dramatic landforms and abundant natural resources. Examination of the physical geography of the region shows processes that continue to shape the diverse landscapes, climates, and vegetation of Africa south of the Sahara. The Sahara Desert is a natural boundary between North and sub-Saharan Africa. North Africa is largely dry and barren with rocky plateaus.

From the top of Shifting Sands dune in the Serengeti Plain of Africa a million mammals are in motion. Wildebeests. Zebras. Gazelles. The plain is black with them. . . . Many of the giant bearded antelope [wildebeests] have newborns trailing them. . . . From the distance the movement seems a serene and constant march toward the southeast where recent rains have made pastures greener.

—Rick Gore, "The Rise of Mammals,"
National Geographic (April 2003)

Land

Africa formed millions of years ago by various physical processes that continue to shape the region today. Africa's shape has changed very little since the continent was formed roughly 225 million years ago. When land masses began to break apart from Pangaea, Earth's original, sole continent, Africa did not move.

Landforms

Africa south of the Sahara is a region formed by shifting tectonic plates. There are two major land types in Africa south of the Sahara: highlands are in the south and east, and lowlands in the north and west. More than half of Africa is desert or dry land.

Mountains and Plateaus

Sub-Saharan Africa has plateaus that rise in elevation from the coast inland and from west to east. They range from 500 feet in the west to 8,000 feet in the east. These plateaus are outcroppings of solid rock that make up most of the continent.

Sub-Saharan Africa has step-like plateaus in the south and east, rising to mountains that are cut in the east by a rift valley. The edges of the plateaus are marked by steep, jagged cliffs called **escarpments**. Most escarpments are less than 20 miles from the coast.

TIP

Rivers crossing the African plateaus plunge down escarpments in cataracts or waterfalls.

Africa south of the Sahara has few mountain ranges. They are not very long and not very high. Small ranges like the Wemmershoek Mountains can be found in Southern Africa.

The Atlas Mountains span 1,200 miles of the North African coast, from Morocco to Tunisia. The tallest of its peaks are in Morocco.

South of the Atlas Mountains, the dry Sahara Desert covers 90% of North Africa.

Eastern Highlands and Mountains

Most African mountains dot the Eastern Highlands, which extend from Ethiopia south almost to the Cape of Good Hope.

The Eastern Highlands include the Ethiopian Highlands as well as volcanic summits like Kilimanjaro and Mount Kenya.

DID YOU KNOW?

Kilimanjaro is the tallest mountain in Africa at 14,298 feet. It is the highest freestanding mountain in the world.

The Ruwenzori Mountains west of the Eastern Highlands divide Uganda and the Democratic Republic of Congo. The snow-covered Ruwenzoris are called the "Mountains of the Moon." Moist Indian Ocean air creates clouds that wrap around the Ruwenzoris.

Further south in South Africa and Lesotho is the Drakensberg range. The Drakensberg Mountains rise to more than 11,000 feet and form part of the escarpment along the southern edge of the continent.

Great Rift Valley

Two deep cuts in the land run from north to south in the eastern part of Africa south of the Sahara. These cuts are called the Great Rift Valley and the Western Rift Valley.

These deep depressions show the effects of two of Earth's tectonic plates pulling apart. Volcanic mountains are located nearby.

DID YOU KNOW?

At its deepest point, the Great Rift Valley reaches nearly 10,000 feet below sea level. It continues to grow larger as Earth's tectonic plates keep pulling apart.

The Great Rift Valley stretches from Syria in Southwest Asia to Mozambique in southeastern Africa. A **rift valley** is a large depression in Earth's surface formed by shifting tectonic plates.

In East Africa, the Great Rift Valley has two branches, with volcanic mountains rising at the edges as well as deep lakes running parallel to its length.

DID YOU KNOW?

Mount Kilimanjaro, on the eastern branch of the Great Rift Valley, has three volcanic cones—Kibo, Mawenzi, and Shira. They are all inactive.

Water Systems

Landforms and physical processes have influenced the water systems of Africa south of the Sahara. These water systems include deep lakes, spectacular waterfalls, and great rivers.

Lakes and rivers drain into huge basins formed millions of years ago by the uplifting of the land. The great rivers of Africa have their sources in the high plateaus and empty into the sea.

Escarpments and ridges break river paths to the ocean with plunging rapids and cataracts. The broken landscape makes it impossible to navigate most of the region's rivers from mouth to source.

Land of Lakes

Most of Africa's lakes are near the Great Rift Valley. Many low areas of the eastern rifts have filled with water, forming a chain of lakes. Located between the eastern and western branches of the Great Rift Valley is Lake Victoria, Africa's largest lake.

DID YOU KNOW?

Lake Victoria is the world's second-largest freshwater lake, second to Lake Superior in North America.

Lake Victoria is the source of the White Nile River, one of the main tributaries of the Nile River, and is only 270 feet deep.

Outside the Great Rift Valley in West Central Africa is the **landlocked** Lake Chad. Lake Chad is fed by three streams but is shrinking due to droughts in the 1970s, water evaporation, and seepage into the ground. Global warming, poor land use and irrigation, and desertification compound the lake's troubles.

DID YOU KNOW?

Through desertification, the desert expands on the dry bottom of Lake Chad.

Lake Tanganyika, one of the deepest and longest freshwater lakes in the world, lies on the western branch of the Great Rift Valley.

South of the Rift Valley is Lake Malawi, a mountain-rimmed lake that resembles a fjord. Fourteen rivers feed into Lake Malawi, and only one flows out from it.

Human-Made Lake—Lake Volta

The damming of the Volta River south of Ajena in the 1960s flooded 700 villages and forced 70,000 people to find a new home. It also created one of the world's largest human-made lakes in the world, Lake Volta in Ghana, West Africa.

The dam was intended to provide hydroelectric power to an aluminum plant. Today, the lake provides Ghana with electricity, supplies irrigation for farming in the plains below the dam, and is stocked with fish.

River Basins

The Nile is the world's longest river, flowing roughly 4,000 miles from East Africa to the Mediterranean Sea. The drainage basin of the Nile covers nearly 10% of Africa. The Nile and its tributaries flow though nine countries: Uganda, Sudan, Egypt, Ethiopia, Zaire, Kenya, Tanzania, Rwanda, and Burundi. It has been used to transport goods since the earliest days of Egyptian civilization.

The Niger River has many names, all signifying "great river." The Niger River is West Africa's main artery. The 2,600-mile-long Niger River originates in the highlands of Guinea. In its great arc, the Niger flows northeast and then curves southeast to the Nigerian coast. The Niger River is used for agriculture and transportation.

In southern Nigeria, the Niger splits into a vast inland **delta**—a triangular section of land formed by sand and silt carried downriver. The Niger Delta stretches 150 miles north to south and is 200 miles wide along the Gulf of Guinea.

Djenné Mali sits along the banks of the Bani River, one of the tributaries of the Niger River that winds through much of West Africa.

South Central Africa's 2,200-mile-long Zambezi River flows from its western source near the Zambia-Angola border to the Indian Ocean in the east where it fans out into a 37-mile-wide delta.

The region's major rivers move quickly through rapids and waterfalls or broaden into marshy inland deltas in lowland areas. The Zambezi's course to the sea is interrupted in many places by waterfalls

At Victoria Falls, on the border of Zambia and Zimbabwe, the Zambezi River plummets 355 feet straight down.

DID YOU KNOW?

Victoria Falls pumps out 500 million gallons of water per minute, about three times as much as Niagara Falls.

The Congo River is different than most African rivers. It reaches the sea through a deep estuary where freshwater meets seawater. The Congo's estuary is 7 miles wide, and ships can navigate the deep water. Smaller boats can navigate the other 2,700 miles through tropical wet rain forests that make a network of the Congo River.

Some parts of the Congo River have rapids and waterfalls that present serious obstacles to traffic.

DID YOU KNOW?

Not far from the Atlantic Ocean, in numerous **cataracts**, the Congo River plunges over 900 feet.

Natural Resources

Mineral resources are among the most abundant natural resources in Africa south of the Sahara. They are not evenly distributed among countries. Metals like copper, iron ore, manganese, and zinc are mined in Africa. Africa is rich in precious gemstones, coal, and metals including gold, iron, and uranium. South Africa, Botswana, and the Congo River basin have major diamond deposits. Nigeria, Angola, Gabon, Congo, and Cameroon have about 4% of the world's oil reserves.

DID YOU KNOW?

South Africa has about half the world's gold and abundant stocks of uranium.

Water is abundant in parts of Africa. West Africa near the equator receives abundant rainfall. Because rainfall is irregular and unpredictable, controlling water for practical uses like irrigation and hydroelectric power is difficult. With a lack of financial support, there is unused hydroelectric power in parts of the region.

The fertile lands of North Africa produce crops such as wheat, oats, vegetables, and citrus fruits, many of which are imported to the United States.

Solar power is a renewable energy source that has been harnessed in the region. A rural electrification program in Kenya has resulted in the installation of more than 20,000 small-scale solar power systems from 1986 to 1996. Solar power use is expanding in Africa south of the Sahara.

Practice

Fill in the blank with the correct word from the information in the preceding paragraphs.

1. At ___________ Falls, on the border of Zambia and Zimbabwe, the Zambezi River plummets 355 feet straight down.

2. A ___________ is a triangular section of land formed by sand and silt carried downriver.

3. Lake ___________ is the world's second-largest freshwater lake, second to Lake Superior in North America.

4. A ___________ valley is a large depression in Earth's surface formed by shifting tectonic plates.

5. While Africa south of the Sahara has few mountains, the ___________ Mountains span 1,200 miles of North Africa's coastline.

6. The world's longest river, the ___________, flows from East Africa to the Mediterranean Sea.

7. Mount ___________, a volcanic summit, is found in the Eastern Highlands.

8. About half the world's ___________ comes from South Africa.

9. Damming a river created one of the world's largest human-made lakes in the world, Lake ___________ in Ghana, West Africa

10. The ___________ River is West Africa's main artery.

Climate and Vegetation

Location near the equator, elevation, rainfall, and ocean air masses influence the climate and vegetation of Africa south of the Sahara. In many places in Africa, water and rain are the same resource.

> **DID YOU KNOW?**
>
> Approximately 1.5 million wildebeests and zebras migrate across the Serengeti Plain every year. Their course is dependent on the rains in Tanzania.

Tropical Climates

Most of Africa lies in the Tropics, so it has tropical climate and vegetation.

Ocean currents, prevailing wind patterns, elevation, and latitude do, however, cause great variations in climate and vegetation.

Tropical Wet

In the tropical wet climate zone near the equator, it is warm, with more than 60 inches of rainfall.

DID YOU KNOW?

The dense tropical rain forests, with an astounding number and variety of life forms, do not have a dry season.

Shrubs, ferns, and mosses grow at the lowest levels of the rain forest. A layer of trees and palms reaches as high as 60 feet above the undergrowth. A **canopy** of leafy trees reaches over 150 feet tall. Orchids, ferns, and mosses grow among branches of the canopy. Tangled, woody vines link trees.

Rain forest soils are not very fertile because heavy rains **leach**, or dissolve and carry, nutrients away from the soil. Bananas, pineapples, cocoa, tea, coffee, and cotton are grown as cash crops on large plantations.

DID YOU KNOW?

The rain forest is threatened by commercial logging and farmers clearing land for these cash crops.

Tropical Dry

Tropical grasslands with scattered trees, called **savanna**, cover about half of Africa south of the Sahara. In the tropical dry climate, there are alternating wet and dry seasons. The wettest areas, closest to the equator, have a six-month wet season followed by a six-month dry season and receive 35 to 45 inches of rainfall a year.

Hot, dry, dusty air streams from the Sahara called **harmattans** arrive on northeast trade winds and soak up the moisture from the summer rains. Cool humid air blows in at the same time from the southwest.

In some parts of the savanna **biome**, trees are the main feature. In other areas, tall grasses dominate. The tropical savanna here and elsewhere is not very fertile.

One of the world's largest savanna plains, the Serengeti, lies in the north central region of Tanzania. It is home to millions of animals; zebras, gazelles, hyenas, lions, giraffes, and cheetahs roam the plain, much of which is now part of Serengeti National Park, which serves to protect the wildlife and natural environment.

Dry Climates

Dry climates of Africa south of the Sahara are located in the north and south. Away from the equator, tropical climates become semi-arid steppe areas and then, the driest climate of all, deserts.

Steppe

In Africa, the semi-arid steppe is a transition zone separating the tropical dry savanna from the desert. The southern transition zone extends to the continent's southern tip.

DID YOU KNOW?

In the north, the Sahel, or "edge" in Arabic, is the band of dry land, or steppe, extending from Senegal to Sudan.

The Sahel has pastures of low-growing grasses, shrubs, and acacia trees. Only 4 to 8 inches of rain fall from June through August. The rest of the year is very dry.

Desertification

The Sahel has lost much land to the desert over the past 50 years. Some scientists believe this **desertification** is due to climate change that brings long periods of extreme dryness and water shortages. Others believe human land use and animal activities have contributed to desertification.

Land is depleted and topsoil eroded by humans stripping trees and animals overgrazing short grasses. Such activities reduce land productivity and the ability to recover from drought.

By 2000, all African countries joined the United Nations Convention to Combat Desertification. The Convention works to have laws enacted that will protect the environment and provide sustainable development.

Desert

The Sahara Desert covers more than 3 million square miles in North Africa. Its boundaries are the Red Sea in the east, the Atlantic Ocean in the west, the Mediterranean Sea in the north, and the Sahel in the south. Mostly undeveloped, the Sahara's landscape is largely shaped by wind over long periods of time, and features ergs, or seas of sand, and hammada, or rocky plateaus.

DID YOU KNOW?

The Sahara isn't only the largest desert in the world, it's also the driest and hottest, too. Temperatures rise above 120°F in the warmest months.

Isolated parts of southern Africa swelter in a desert climate. In the east, hot, dry weather prevails in most of Kenya and Somalia. The Namib Desert along the Atlantic Coast of Namibia in southern Africa has rocks, dunes, and sparse desert plants.

The Kalahari Desert occupies the interior of eastern Namibia, most of Botswana, and part of South Africa. The Kalahari is sand-swept and very dry. Parts of it support some animals and a variety of plants, including grasses and trees.

In the Kalahari Desert, average monthly temperatures are high and there is little rainfall. Daily temperatures in the Kalahari range from 50°F at night to 120°F during the day.

DID YOU KNOW?

Winter in the Kalahari Desert brings even colder or freezing temperatures at night.

Midlatitude Climates

Not as extensive as tropical and dry climates, midlatitude climates exist in Africa south of the Sahara. The southern coastal areas of South Africa are characterized by marine and humid subtropical climates.

The East African highlands have comfortable temperatures and enough rainfall for farming. The lush green vegetation of protected forests and farm crops abounds and snow even falls in the Highlands.

Vertical climate zones exist in the Highlands of Africa, just as they do in Latin America. Temperature decreases as elevation increases. Lower elevations support woodlands and agriculture, while higher elevations support shrubs and some grasses.

Practice

Insert T or F to indicate whether the statement is true or false, based on what you've read.

11. _____ Heavy rains keep the soil of the rain forests fertile.

12. _____ Bananas, pineapples, cocoa, tea, coffee, and cotton are grown for subsistence, not profit.

13. _____ The savanna is very fertile.

14. _____ Hot, dry, dusty air streams from the Sahara are called trade winds.

15. _____ Most of Africa lies in the Tropics.

16. _____ Tropical rain forests in Africa do not have a dry season.

17. _____ Tanzania's plains have been fed by rain; there is a great migration on the Serengeti Plain.

18. _____ The Sahel has a tropical wet climate.

19. _____ The area around Cape Town has a subarctic climate.

20. _____ Vertical climate zones exist in the Highlands of Africa.

Answers

1. Victoria
2. Delta
3. Victoria
4. Rift
5. Atlas
6. Nile
7. Kilimanjaro or Kenya
8. Gold
9. Volta
10. Niger
11. F
12. F
13. F
14. F
15. T
16. T
17. F
18. F
19. F
20. T

LESSON

16 AFRICA—HUMAN GEOGRAPHY (PART I)

LESSON SUMMARY

Current events in Africa are best understood by knowing about the region's thousands of diverse ethnic groups with different histories, languages, religions, and cultures.

Maps of the ancient sea kings demonstrate Africans along with many Middle Eastern peoples came to America way before the Vikings. After all, if the Garden of Eden, the birthplace of humanity is anywhere, it is in Africa.

—John Henrik Clarke, *Introduction to African Civilizations*

Countries

Angola
Benin
Botswana
Burkina Faso
Burundi
Cameroon
Cape Verde
Central African Republic
Chad
Comoros
Congo, Democratic Republic of the
Congo, Republic of the
Cote D'Ivoire
Djibouti
Equatorial Guinea
Eritrea
Ethiopia
Gabon
Gambia
Ghana
Guinea
Guinea-Bissau
Kenya
Lesotho
Liberia
Madagascar
Malawi
Mali
Mauritania

Mauritius
Mozambique
Namibia
Niger
Nigeria
Rwanda
São Tomé and Príncipe
Senegal
Seychelles
Sierra Leone
Somalia
South Africa
South Sudan
Sudan, Republic of
Swaziland
Tanzania
Togo
Uganda
Zambia
Zimbabwe

Less commonly referred to as African, these Northern countries also share the African continent: Algeria, Egypt, Libya, Morocco, Sudan, Tunisia, and Western Sahara.

The Sahel

The Sahel, a transitional zone between the Sahara and the Sudan region, influences ways of life in the subregion. It covers parts of Senegal, Mauritania, Mali, Burkina Faso, Algeria, Niger, Nigeria, Chad, Sudan, Somalia, Ethiopia, and Eritrea.

Population Patterns

The changing physical environment and many diverse ethnic groups have shaped population patterns in the Sahel. The farmers and herdsmen of various ethnic groups who live off the land have endured drought, deforestation, and overpopulation.

People

Hundreds of ethnic groups coexist, influenced by Arab, European, and indigenous African cultures. Major groups include the Mandé peoples of Senegal and Mali, the Hausa of Niger, and the Wolof of Senegal. Many people in the Sahel follow other traditional African religious and cultural practices.

Density and Distribution

The Sahel has rapid population growth, but low population density spread unevenly across the subregion. In the Sudan, virtually all people live along the Nile River. Areas of Mali and Mauritania are also uninhabited.

Increased desertification and deforestation of the Sahel have pushed people into the cities of Dakar in Senegal, Niamey in Niger, and Bamako in Mali. Senegal is the most urbanized country, with 43% of people living in heavily European-influenced Dakar.

The physical environment and relative location of the Sahel have brought together diverse cultures that continue to influence the subregion. The Sahel's location near Europe and Southwest Asia has made it susceptible to numerous invasions and migrations.

History and Government

First Civilizations

The fertile Nile River Valley gave birth to Egyptian civilization. Between 2000 and 1000 B.C., the Egyptians gained control of cultures along the Nile.

Empires and Colonization

Trading empires dominated West Africa. The gold-for-salt trade started in the empire of Ghana was taken up by the Mali and Songhai Empires. Extending west to the Atlantic Ocean, Mali's wealthy city center was Timbuktu.

The wealth of the African kingdoms reached medieval Europe. Europeans brought back gold and African goods as early as the 1200s. By the 1600s, European trade with Africa for gold, goods, and slaves was extensive. In the 1800s, European powers treated the Sahel as a source of raw materials and a market for finished goods. The entire Sahel region was under European control by 1914.

Sudan Today

There has been ongoing conflict in the Sudanese region of Darfur since 2003. On one side are Arab Muslims in the north who are characterized as intellectuals, and on the other, non-Arab Muslims from the south who are looked down on and called tribesmen and slaves. These racist attitudes have fueled a genocide that has taken hundreds of thousands of lives, and also brought on starvation, disease, and migrations. After years of civil war, South Sudan gained independence as a nation in July 2011.

Arts

African art often expresses traditional religious beliefs. African visual arts include the ceremonial masks and wooden figures of the Dogon people of Mali. Musical traditions of Africa include percussion, the talking drum, and five-string guitar. Oral communication has a strong tradition in the Sahel.

Written African literature developed mostly in northeast Africa from contact with early Mediterranean writing systems. Nafissatou Nian Diallo's 1975 autobiography was one of the first literary works published by a woman from Senegal.

Family Life

In rural areas, people often live in extended families that are patriarchal, or male-dominated, with female support. People are organized into **clans**, or large groups of people descended from an early common ancestor. In the cities, the **nuclear family** made up of husband, wife, and children is replacing the extended family.

Language and Religion

By some estimates, more than 2,000 languages are spoken in Africa. In the Sahel, there are Afro-Asiatic, Nilo-Saharan, and Congo-Kordofanian language groups. Nilo-Saharan speakers, for example, live in southern Sudan and Chad. Due to a legacy of colonial rule, French is spoken throughout Africa.

Islam is the main religion of the Sahel. Christianity is practiced by many people in Chad, Sudan, Senegal, and Niger. Many people of the Sahel retain their indigenous religious practices, including rituals involving a supreme being and lesser-ranked deities.

Education and Healthcare

School enrollment and literacy rates are low in the Sahel. In the poorest countries of Niger and Mali, less than a third of children attend school.

Poverty in the Sahel results in high infectious disease and mortality rates. The Sahel lacks adequate healthcare during pregnancy and childbirth. The subregion has high infant and maternal mortality rates. Very few rural Africans have access to clean water.

East Africa

East Africa's peoples, history, and cultures have been influenced by its location near the Indian Ocean, which has been a gateway between the trading ports of Africa, Asia, and the Arabian Peninsula.

The Swahili people living along the East African coast, for example, are descendants of East African, Arab, and Persian traders who made their living in the ports of Zanzibar, Dar es Salaam, and Mogadishu.

Population Patterns

East Africans are as diverse as the subregion's terrain. The people of East Africa live along the coasts, in desert and steppe areas, and in the highlands along the Great Rift Valley.

People

In many East African countries, one ethnic group is the majority. The Bantu make up most of Uganda and Tanzania. The Hutu make up the majority in

Rwanda and Burundi. There are ample Arab and European influences in East Africa.

Density and Distribution

Due to the land and climate, population distribution is uneven in East Africa. In Tanzania, for example, population distribution ranges from 3 to 133 people per square mile.

Most cities lie on coasts or rivers, but some inland cities—like Nairobi in Kenya and Addis Ababa in Ethiopia—grew because of trade.

DID YOU KNOW?

About 60% of people in Somalia are nomadic or seminomadic.

Farmers are producing less in East Africa while population growth soars rapidly. East African governments have pushed the export of cash crops to boost national incomes. Meanwhile, not enough food has been produced for domestic needs. Poor farming practices have exhausted huge expanses of farmland, and drought makes the situation worse.

History and Government

Throughout most of its history, East Africa's location has attracted people from many continents. East Africa was home to the world's first humans, countless indigenous peoples, and various European colonizers.

Early Peoples and Kingdoms

Archaeologists have discovered human bones dating back 3.2 million years in Ethiopia. Human bones over 2.6 million years old have been found in Kenya.

Arabs first settled East Africa in the A.D. 700s. They brought the Arabic language and culture with them. Persians settled the East African coastline from about A.D. 700 until the Portuguese claimed control in the late 1400s.

East Africa's location along the Red Sea coast near the Arabian Peninsula gave it trading relationships with Arabian, Asian, and Mediterranean civilizations.

European Colonization

The Portuguese who explored the East African coast brought Roman Catholicism to Ethiopia. Arab dominance in East Africa declined and hatred toward Europeans and Christians lasted into the 1900s.

New steam-powered transportation and disease-treating medicines allowed Europeans to explore the mysterious "Dark Continent." British doctor and missionary David Livingstone was one of Africa's first explorers; he challenged Europeans to spread commerce, Christianity, and their culture across Africa.

European powers—mostly Britain, France, Germany, and Portugal—competed ferociously to expand their empires and protect trade routes. In 40-year periods, Europeans carved out of Africa more than 40 countries.

Colonies to Countries

Most African groups who resisted foreign rule in the colonial period failed. Independence movements were more successful in the 1950s. Increasing pressure gave many African countries self-rule in the 1960s, resulting in internal strife.

In Uganda, a brutal dictatorship under Idi Amin in the 1970s caused social disintegration, human rights violations, and economic decline.

In Rwanda and Burundi, colonial powers gave Tutsi people positions of power over the Hutu. The Hutu became resentful and beginning in 1959, resorted to violence. The decades-long violence reached its genocidal height in Rwanda in 1994 when the Hutu killed thousands of Tutsi. Somalia and Ethiopia have had border disputes that lasted for years. Warring factions, famine, and drought cause governments to collapse or weaken.

International efforts to halt violent conflicts often fail. Although local courts in Rwanda are trying suspects accused of genocide, violence and instability prevails in East Africa.

Language and Religion

East Africa has been influenced by the languages, religions, and ways of life of its colonizers. As in the Sahel, most East African languages fit into one of the three major language groups: the Congo-Kordofanian, Nilo-Saharan, or Afro-Asiatic.

Arab settlers brought the language and religion of Islam to the northern part of East Africa. Arabic is the most common language in Sudan and Eritrea. Indigenous languages and religions, with some Arabic and European influences, dominate further south in East Africa.

Christianity was adopted in Ethiopia in the A.D. 300s. It did not spread across all of East Africa until the colonial period. Most people in East Africa are Muslim or Christian, but traditional African religions are common as well.

Education and Healthcare

Literacy rates are highest in urban areas and have been going up in rural areas since independence. Literacy rates range from 35% in Ethiopia to 70% in Uganda. Only a small percentage of East Africans complete a secondary education.

Although healthcare is improving in East Africa, many problems remain due to poor nutrition, famine, overpopulation, and the inability to cure common diseases.

Drugs that can control the diseases are available in developed countries, but most African governments and individuals cannot afford to buy them. HIV and AIDS are more widespread in Africa south of the Sahara than anywhere else in the world. AIDS has cut life expectancy throughout East Africa and other parts of sub-Saharan Africa.

Varied Ways of Life

Ways of life in East Africa are as varied as the ethnic groups who live there. Of the 120 ethnic groups in Tanzania, the Sukuma farm land south of Lake Victoria, while the Chaggas grow coffee in plains around Kilimanjaro. In Tanzanian cities like Dar es Salaam and the capital Dodoma, people work in factories and offices.

In urban areas, families live in high-rises with modern conveniences. Rural farmers often live in thatched roof dwellings without modern or sanitary conveniences.

The colorful Masai peoples of Kenya and Tanzania are **pastoralists** who do not farm. They live in **hierarchically** organized settlements in which elders make important decisions.

West Africa

West Africa's religions and social structures play a significant role in people's daily lives.

Population Patterns

The locations and densities of West African populations affect people's way of life. With increasing population, decreasing food supplies, deforestation, and climate change, West Africans from Benin to Togo to Côte d'Ivoire to Guinea are moving to urban areas seeking jobs and education.

People

There is a wide variety of ethnic groups in West Africa. Some groups have lived there for centuries. Some ethnic groups were split apart by colonial European boundaries and live in adjoining countries. The Hausa, for example, who lived along the caravan trade route to Southwest Asia and were traders and farmers, live in northern Nigeria and southern Niger.

The Yoruba are one of the largest ethnic groups in Africa. The Yoruba live in the grasslands, forests, and cities of Togo, Benin, and southwest Nigeria.

DID YOU KNOW?

Over 20 million people in West Africa speak the Yoruba language, which is part of the Congo-Kordofanian language group.

Density and Distribution

West African countries face rapid population growth. Nigeria's population is expected to rise from 150 million in 2011 to 390 million by 2050. Increasing population growth in Niger is causing competition between herders and farmers for natural resources and land.

Most people in West Africa live along river plains and the coast because of the fertile soil and mild climates that have drawn agriculture, industry, and commerce there.

Although most Africans still live in rural areas, as population growth and climate change deplete natural resources, West Africans are moving from rural areas to urban settings for better job opportunities, healthcare, and public services. Cities have spread into the countryside. Towns and villages have become service centers for rural dwellers, who travel there by foot, bus, or boat.

Over 60% of people in Gambia live in villages. Half of Senegal is rural. Despite Nigeria's rapid urban growth, only 44% of its people live in cities.

History and Government

Indigenous and outside forces with their own cultures shaped West African history. The resources that built empires attracted wealth-seeking European powers. Powerful West African empires were replaced by colonial rule and later dissolved into independent countries. Colonial rule and ongoing economic problems have left West Africa with many challenges.

Early Empires

West African empires grew strong by virtue of trade around the A.D. 700s. The modern countries of Ghana and Mali are named after these ancient empires.

Ghana grew wealthy trading gold for salt brought by camels across the Sahara. Ghana prospered for 500 years and its power was reflected in its large capital of Kumbi.

Colonial Era

By the 1400s, the gold-seeking Portuguese had set up trading posts along the African coast. Foreigners who saw the trading centers of Timbuktu, Kano, Gao, and Wangara were impressed with the bustling markets and thriving cultures.

African kings had enslaved and traded prisoners of war for centuries. Arab traders brought enslaved Africans to the Muslim world starting in the A.D. 800s. The slave trade increased in the 1600s and 1700s.

Colonial Legacy: Nigeria

The British formed the colony of Nigeria in 1914, combining several small ethnic territories. In the north, the cultures were based on Islam. In the south, the cultures were based on African religions or Christianity. After independence in 1960, ethnic and religious difficulties resulted in civil war in Nigeria. Such ethnic and religious divisions trouble Nigeria today.

Language and Religion

Hundreds of languages are spoken in West Africa. There are over 250 languages and cultures in Nigeria alone. The language of instruction and official language of Nigeria is English, but most people do not use it regularly. The Congo-Kordofanian language of Yoruba is printed in books, newspapers, and pamphlets and taught in radio and broadcasting schools, secondary schools, and the universities of the southern part of West Africa. English and French are widely spoken and Arabic is common in the northern areas of West Africa.

Religion is an important part of everyday life in West Africa. Islam, Christianity, and traditional African religions are the most common religions in West Africa. The different religions generally live peacefully, but conflict does sometimes occur between competing religious groups.

Education and Healthcare

Free and universal education is inconsistent in West Africa. Literacy rates rang from 18% in Niger to 75% in Ghana, where the constitution mandates mandatory and free primary and secondary education. Students in Ghana who complete high school attend universities or polytechnic schools to learn trades. Sierra Leone's support for education has declined over the past 30 years, whereas over 4 million students attend school in Ghana today.

Arts

Most West African art expresses religious beliefs. Music and dance are part of everyday life. Whole communities participate while dancers wear masks honoring deities or ancestral spirits, or celebrating births.

DID YOU KNOW?

Gospel, jazz, blues, ragtime, rock and roll, hip-hop, and rap all owe great debts to the West African musical style of "call and answer" brought to the Americas by enslaved Africans.

The ways of life of Africans are connected to their arts. The Ashanti of Ghana are expert weavers known for the kente cloth worn by African rulers for centuries. The cloth is now a symbol of Africa for African Americans. Nigerian art includes wooden masks used in religious ceremonies and sculptures used as objects of worship.

Central Africa

Rain forests cover more than half of Central Africa. People living in the dense forest must contend with different growing conditions than those in more open areas. It is a natural environment unlike any other in Africa. While indigenous peoples built societies in response to the natural environment, European powers exploited the region.

Population Patterns

Groups of people in Central Africa share a culture based on physical environment. Central Africans live in a temperate climate with thick tropical forests and tropical grasslands. Central Africa is mostly rural with few urban areas.

People

Dense natural vegetation near the equator makes large-scale farming difficult, so most people are subsistence farmers or cattle herders. In some places, cacao and coffee are grown for export.

Like the rest of Africa, Central Africa has hundreds of ethnic groups. At least 250 groups live in the Democratic Republic of the Congo and Cameroon. In Cameroon, peoples of the north such as the Fulani are predominantly Muslim.

There are more than 40 ethnic groups in Gabon who speak different languages. The Fang group of Equatorial Guinea is divided into 67 clans of Bantu origin.

The Mbuti of Central Africa live in the forest, unchanged by outside influences. The Mbuti are hunters and gatherers who live off animals and fruits.

The 145,000 inhabitants of the islands of São Tomé and Príncipe descend from Portuguese settlers and freed enslaved people who migrated there since the late 1400s.

Density and Distribution

Central Africa is one of the least densely populated regions of Africa. Gabon has so few people it still has a labor shortage despite a recent surge in immigration.

The Republic of the Congo, the Democratic Republic of the Congo, and Cameroon are the more densely populated parts of Central Africa. Although the northern part of the Democratic Republic of the Congo is minimally inhabited tropical forest, the rest of the country, including its capital and the region's political, cultural, and economic hub, Kinshasa, holds millions of people.

> **DID YOU KNOW?**
>
> The western highlands of Cameroon are believed to be the starting place for the massive Bantu migrations across Africa 2,000 years ago.

The west is the most densely populated part of Cameroon, where people engage in intensive agriculture and commerce.

History and Government

Central Africa was a location of constant early migrations, and later, of European systems of slavery, colonization, and plantation economy. When Europeans arrived, Central Africa already had huge indigenous trading empires.

Early Settlement

Although people lived in Central Africa for more than 10,000 years, large numbers of African people did not arrive until the A.D. 600s.

The Bantu established the Central African kingdoms of the Congo, Luba, and Luanda which are today Tanzania, Malawi, Zambia, and Zimbabwe.

Slavery

European explorers came to the coasts of Central Africa in the 1400s to trade. They carried away slaves from the areas that became known as the Republic of the Congo and Cameroon. Gabon became a slave-trading center. As many as 10 to 20 million people from the African interior were sold into slavery. King of the Congo Nzinga Mbemba complained to the king of Portugal that slave merchants were utterly depopulating his country.

European Colonization

Large parts of Central Africa remained uncolonized until the 1800s. France made treaties with local rulers in the Republic of the Congo to protect them from Belgium, France's main rival. The French directed the Congo economy toward resource extraction and cash-crop farming for export. Towns rose along the railway line to transport goods to the coast.

Huge plantations and cash crops disrupted village life and local agriculture. Eventually, African culture was destroyed by European rulers, businessmen, missionaries, and culture.

By the early 1900s, there were frequent revolts against French rule and plantation economy. Resistance to colonial rule increased after the mid-century. After Central Africans fought for France in World Wars I and II, France instituted reforms in Central Africa. By 1960, all French colonies were independent.

Post-Independence Instability

Following independence, most Central African countries experienced ethnic strife, human rights abuses, and harsh rule. In the Democratic Republic of the Congo, serious instability led to the reign of the dictator Mobutu Sese Seko from the 1960s to the 1990s. This period saw one-party rule, human rights abuses, and war.

Due to Central Africa's abundant natural resources, some countries in the region have achieved economic stability. Increasing oil revenues have helped countries like the Republic of the Congo ease debt burdens, but many countries still face unrest and instability.

Language and the Arts

Central Africa has hundreds of languages. Over 700 languages are spoken in the Democratic Republic of the Congo. A simplified form of speech called **pidgin** is used by people who speak different languages.

The Mangbetu are world renowned for beautiful pottery, sculpture, and building. Sculptures depicting elongated heads are treasured by the Mangbetu.

> **DID YOU KNOW?**
>
> French colonial rule made French a regional language throughout Central Africa.

Religion and Family Life

Religion and family life intertwine in Africa. In the Democratic Republic of the Congo, some ethnic groups mix indigenous and Christian beliefs. Labor is frequently divided by gender. Many generations and even people from different families live in the same dwelling.

Traditional religions in Central Africa are numerous and diverse, but have common elements. They have a supreme being and a ranked order of deities, and practitioners believe in nature spirits and honor distant ancestors and family members who have died.

European colonialism widely impacted religious practices in Central Africa. French colonists and missionaries brought Roman Catholicism to Central Africa. In many areas, a majority of people practice a form of Christianity.

Education and Healthcare

Religion dominated education during colonial rule, but violent conflicts and economic problems have hurt Central Africa's educational systems. There have been increased efforts in education since independence, but many rural areas still do not have schools. Literacy ranges from 50% in the Central African Republic to about 85% in Equatorial Guinea. Women have a much lower literacy rate.

Equatorial Guinea achieved high literacy and a fairly good healthcare system under Spanish colonial rule.

Most Central African countries do not have the resources to prevent diseases Western countries have stopped. Central Africa's primary health concerns are a lack of safe drinking water, a shortage of vaccines for curable diseases, and the rising number of AIDS victims.

Practice

1. In the sixteenth century, the ___________ colonized Angola and Mozambique.

2. People living in the dense ___________ ___________ that cover more than half of Central Africa must contend with different growing conditions than those in more open areas.

3. In the Sudan, virtually all people live along the ___________ River.

4. The ___________ peoples of Kenya and Tanzania are pastoralists who do not farm.

5. ___________ settlers brought the language and religion of Islam to the northern part of East Africa.

6. In Rwanda and Burundi, colonial powers gave ___________ people positions of power over the Hutu.

7. The poorest countries of the Sahel are ___________ and ___________.

8. The ___________ people living along the East African coast are descendants of East African, Arab, and Persian traders.

9. Colonial rule made ___________ a regional language throughout Central Africa.

10. After ___________, many European colonies in Africa won self-rule.

Answers

1. Portuguese
2. Rain forests
3. Nile
4. Masai
5. Arab
6. Tutsi
7. Niger and Mali
8. Swahili
9. French
10. 1950

LESSON

17 AFRICA—HUMAN GEOGRAPHY (PART II)

LESSON SUMMARY

This lesson continues the rich history of Africa south of the Sahara, touches on its relationship with the United States, and discusses environmental and other problems it currently faces.

I dream of the realization of the unity of Africa, whereby its leaders combine in their efforts to solve the problems of this continent. I dream of our vast deserts, of our forests, of all our great wildernesses.

—Nelson Mandela

Population Patterns

Southern Africa is experiencing significant population changes as it overcomes its colonial history and confronts present challenges.

People

Throughout southern Africa, political borders have not stopped ethnic, linguistic, religious, and cultural overlap. Ethnic groups often share communal and family traditions.

The Sena live in the marshes near the Zambezi River dividing Zambia and Zimbabwe. They travel by canoe up the Zambezi River to sell fish. They buy items in markets in Malawi. They float downriver to Mozambique to trade fish for sugar.

The Bantu ethnic group survives all over southern Africa. The Swazis migrated to what is now Swaziland in the 1500s and had to contend with the Zulu population that has remained strong in what is now South Africa.

The nearly culturally extinct San Bushmen live remotely on the edges of the Kalahari Desert of Botswana, Namibia, Angola, South Africa, Zimbabwe, and Zambia. The San are a link to the hunter-gatherer way of life that prevailed until 10,000 years ago.

Density and Distribution

Population densities vary in southern Africa. Namibia has 6 people per square mile and Lesotho, surrounded by South Africa, has 106 people per square mile.

Although some people practice **subsistence farming** and herding in southern Africa, many are moving to urban areas for jobs in gold and diamond mines. Johannesburg, South Africa, grew because of gold mining.

There are skyscrapers and trendy shopping centers in the cities of southern Africa. Most city residents, though, face inadequate public services, overcrowded neighborhoods, and pollution. Over 35% of Zambia is urban. In the 1960s, white minority land reforms in Zimbabwe forced black Africans off farms to live in crowded areas around cities.

Population explosion in parts of Africa is counterbalanced by deaths from AIDS. Over 70% of the 36 million people carrying HIV live in Africa south of the Sahara. Over 2 million people in the region died of AIDS-related diseases in 2005.

DID YOU KNOW?

The United Nations estimated in 2010, 10.7 million children under 15 years old lost at least one parent to AIDS.

History and Government

Situations created by colonial rule challenge southern Africa today and will continue to do so in the future.

Early Cultures

Cultural artifacts over 1 million years old have been found in southern Africa. Stone ruins of the "Great Zimbabwe" date back to A.D. 800–1200. These indigenous people traded along the southeastern coast of Africa.

The Zulu culture is one of Africa's oldest. The Zulu, like many Africans, descend from the Bantu who migrated across southern Africa.

European Colonization

People in Madagascar were trading with outsiders starting in the A.D. 600s. Arabs and Europeans settled in southern Africa for economic reasons. They set up trading posts and highly restrictive colonies. Starting in the late 1700s, ships carried rice from Madagascar to the new state of South Carolina.

In the 1500s, the Portuguese set up a strong slave trade in Angola. Workers and slaves were sent to other Portuguese colonies and the Americas. In Mozambique, the Portuguese built railroads to transport labor to their mines and plantations. The Portuguese organized trading colonies in Comoros and Mauritius.

Africans did not benefit and were often exploited by outsiders across Africa. Such practices created conflict and independence movements across the continent.

Challenges Post-Independence

Transition to independence was not easy in most southern African colonies. Unrest grew in the early to mid-1900s. In Angola, there was a quick **coup d'état**, or overthrow of the Portuguese government, in 1975.

Post-colonial rule has been difficult for most African countries. Although countries in southern Africa enjoy more freedoms under independence, they face serious economic and healthcare problems.

Vast resources were not enough to rebuild Zambia after independence. Few people were trained to run the government, the economy depended on foreign oversight, and civil wars in neighboring countries pushed refugees into an already overburdened country.

Although South Africa declared independence from British rule in the early 1900s, the white minority ran the government throughout the century by racially separating black Africans and racially mixed peoples in a system called **apartheid**.

> **DID YOU KNOW?**
>
> Under apartheid, the minority white Afrikaners of European descent, who wished race-based slavery to continue, enacted strict codes separating blacks while denying them the right to vote.

In the early 1990s, internal unrest and international pressure forced South Africa to end apartheid. The country's leading antiapartheid activist, Nelson Mandela, was freed after 27 years in prison. In 1994, South Africa held its first election based on **universal suffrage**, or voting rights for all adult citizens. Mandela was elected South Africa's first black president.

South Africa has moved from repression toward democracy, but millions of black South Africans continue to live in poverty and the legacy of inequality remains. South Africa is striving to meet the challenge of a better life for all its citizens.

Botswana has been stable and economically successful since independence in 1966, however, and Mauritius has established an open economy and political system with coalitions and alliance building.

Culture

Non-African languages of southern Africa include the Indo-European languages of English, French, and Afrikaans brought by traders, administrators, and missionaries. Afrikaans was brought by early Dutch settlers and includes German, French, and English words.

Christianity is the most common religion of southern Africa. Traditional religions are also practiced.

Education has expanded in southern Africa since independence. Mozambique is constructing new schools and training teachers to match population increases. Rural areas often lack qualified teachers and resources.

Southern Africa has a wide variety of indigenous art, crafts, music, and dancing. In 2003, South African novelist J.M. Coetzee won the Nobel Prize in Literature. In 2005, the South African film *Totsi* won the American Academy Award for Best Foreign Language Film.

People play games and spend leisure time with family in southern Africa. They embrace local culture, including music and dance. Urban dwellers have access to Western movies and music. They often have Western style clothing, use cell phones, and watch television.

Economy

Africa south of the Sahara is making the slow transition from a subsistence-based farm economy to a participant in the global economy. There are economic imbalances among the African countries south of the Sahara due to an uneven distribution of natural resources.

Agriculture

More than two-thirds of Africans work in agriculture. Some countries produce a single crop and others a variety. Most Africans south of the Sahara engage in subsistence farming to provide for their family or village. Extra harvest or animals are sold at a local market.

Farming Methods and Export Crops

Farmers use a variety of methods to work the land. The Masai of Kenya and Tanzania and the Fulani of Nigeria and other parts of West Africa are pastoralists, meaning they raise livestock. In the forest, farmers use the **slash and burn**, or **shifting cultivation**, farming technique. They burn trees and brush they cut down and then plant seeds in the ash-rich soil. When the soil is no longer fertile, they move and return when the soil has renewed itself.

Sedentary farming occurs at permanent settlements with good soil. The Kikuyu in Kenya and Hausa in Nigeria farm permanent plots. Europeans in South Africa, Kenya, and Zimbabwe also practice sedentary farming

A small percentage of Africans work at large-scale commercial farming. Cash crops are grown and sold for profit.

Commercial farms supply the world with palm oil, peanuts, cacao, and sisal, a vegetable fiber used in rope.

The colonial economic system deeply influenced the growth of commercial farming in Africa. The same crops are the region's main agricultural crops today. Côte d'Ivoire, Nigeria, Ghana, and Cameroon depend on the sale of cacao, used to make cocoa and chocolate. Kenya, Tanzania, and Madagascar produce tea and coffee. Most crops leave Africa to be processed elsewhere, as in colonial times.

Land Conflict: Zimbabwe

Cash crop plantations and large-scale farms take the best land. This makes it hard for farmers to meet their own needs.

Zimbabwe's agriculture-based economy has collapsed. Less than 1% of its farmers are white, but they own 70% of the land. In 2000, Zimbabwe's President Mugabe began redistributing land more evenly. Small-scale farmers tried to take over large farms seized by the government without compensation to the owners. The result has been violence, even though many white farmers support some reform.

Meeting Challenges

Overgrazing, overworked soil, and lack of technology make farming difficult. Frequent tilling, clearing forests, and use of heavy machinery have caused erosion and desertification. Men work mostly in cash crop production while women use basic tools for subsistence farming. Food production has fallen short of the region's needs.

Some people have begun **conservation farming**, or growing crops where they grow best, to save land. Better seeds, irrigation, and fertilizers have also increased crop yields and production.

Logging and Fishing

Logging in Africa only provides about 10% of the world's lumber supply. Coastal rain forest countries like Gabon and Côte d'Ivoire export large amounts of hardwoods such as Rhodesian teak, ebony, African walnut, and rosewood.

Few African countries build and support commercial fishing fleets. Africa has a very narrow continental shelf near the coast, so fishes are scarce there. The southwestern coast does yield sardine, tuna, and herring for export.

Mining Resources

Miners do a dangerous and important job to support their families. The 300-mile-long Witwatersrand gold deposit makes South Africa the world's largest gold producer. South Africa leads the world in the production of gems and industrial diamonds. Its mineral wealth makes it one of the region's richest countries.

Diamonds have been in high demand for thousands of years. Most commercial diamonds come from Africa. The largest deposits of diamonds are in South Africa, Namibia, Botswana, the Democratic Republic of the Congo, Angola, Tanzania, and Sierra Leone.

Most mineral deposits lie along the Atlantic coast south of the equator. Abundant oil reserves make Nigeria the region's only member of the Organization of Petroleum Exporting Countries (OPEC). Oil fields recently discovered off the coast of Equatorial Guinea have jump-started that country's economy.

Despite Africa's mineral wealth, most people never benefit due to government mismanagement of mineral income and foreign corporations' sending their profits abroad. Chad became an oil-producing state in 2003 and, unlike Nigeria, is trying to use oil profits to protect the environment and combat poverty.

Industrialization

Most countries never had the **infrastructure** to develop manufacturing industries, so they often need foreign loans to develop industries. But there are barriers to industrialization in Africa. There is a lack of skilled workers, so education and skills training are essential. There are power shortages, and hydroelectric resources are underutilized.

Since the 1960s, African governments have encouraged industrial expansion. The region's workers produce textiles, processed food, paper goods, leather products, and cement.

Transportation and Communications

Developments in transportation and communications should have positive effects on the region's economy, but the physical environment makes creating and maintaining such systems challenging. Roads and railways must cover vast distances and difficult terrain. Water transportation is limited because some rivers cannot be navigated from source to mouth and the region has few natural harbors.

Roads and Railroads

Nigeria plans to link all its railroads. Uganda is repairing the Trans-Africa Highway, which runs from the urban-industrial centers of Mombasa, Kenya, to Lagos, Nigeria. Mauritania, Senegal, and Morocco in North Africa are finishing a 1,875-mile highway between Tangier, Morocco, and Dakar, Senegal, that will ultimately reach Lagos. The Trans-Sahara Highway that opened in 2003 links markets in different parts of Africa with Europe. Another highway connects Dakar, Senegal, with N'Djamena, Chad.

People and Environment

Human activities like war and deforestation have had a tremendous impact on the environment and have contributed to famine and starvation of millions of people.

Managing Resources

Environmental challenges threaten Africa's food, healthcare, and plant and animal life. In Africa, poverty and hunger threaten millions of people every day.

In Africa south of the Sahara, factors like poverty, population growth, war, and drought have severely strained the environment. Over 31 million people in Africa south of the Sahara are in serious need of food.

In the 1990s, thousands of people died of starvation on the Horn of Africa, the strip of land that juts into the Indian Ocean and includes the countries of Somalia, Ethiopia, and Djibouti. The famine was caused by drought and war.

Desertification

Africa south of the Sahara has many areas with dry climates and poor soils. Thousands of years ago, pastoral peoples and livestock stripped fertile Sahel of its vegetation. The desert is now spreading south into Mauritania, Mali, Niger, Chad, and Sudan. The **carrying capacity**, or number of people a region can sustain, has been greatly exceeded.

Droughts have become severe in the Sahel and other parts of Africa south of the Sahara. Severe droughts in the 1970s turned farmland to wasteland. Since 1998 in East Africa, drought has killed crops and livestock and threatened the lives of millions of people. The 2004 drought in Niger put subsistence farmers through a major food crisis. The next year, 3 million people needed food aid.

In 2000, the United Nations Food and Agriculture Organization (FAO) warned that famine could become a problem in Central Africa because of unpredictable weather patterns and refugees. Good harvests in West Africa have increased food supplies in most countries, but civil war endangers food distribution in Sierra Leone, Liberia, and Guinea.

Conflict and Hunger

Since 1990, conflicts in Liberia, Sudan, Somalia, and Rwanda have stopped economic growth and caused widespread starvation. Massive refugee populations leave war-torn areas, cross borders, and put a strain on already meager food sources.

Somalia has been without a government since 1991 and 2 million people there are threatened by civil war.

Looting and fighting cripples food distribution.

In former Sudan, decades of civil war between the Muslim Arab government in the north and non-Muslim rebels in the south has, along with periodic drought, created the world's largest refugee population. The 2005 peace did nothing about the separate conflict and world's worst humanitarian crisis in western Sudan's Darfur region.

To solve the hunger crisis, peace is necessary. Ethiopia and Eritrea halted their two-year war in 2000. Although one of the worst droughts in the area's history continues, Ethiopia and Eritrea maintain a shaky peace and that gives farmers hope.

Farming in Peace

Since Eritrea became independent of Ethiopia in 1993, farmers have steadfastly worked to improve the land. Farmers in the Ethiopian province of Tigray terraced 250,000 acres of land and planted 42 million trees to hold soil in place. They built earthen dams to store rainwater. Grain crops thrived. In Eritrea, crops were so abundant that its relief requests to other countries were cut in half.

International Red Cross's feeding centers and Doctors Without Borders's medical teams are nursing malnourished children and adults back to health in Africa. Humanitarian aid continues throughout Africa.

Human Impact

Human activities have destroyed rain forests, threatened wildlife, and raised questions about land use in Africa south of the Sahara.

> **DID YOU KNOW?**
>
> In 2000, Africa's tropical forests were disappearing at a rate of 12 million acres each year.

Hunting and tourism also raise serious land use concerns.

Tropical Forests

In 1990, Africa south of the Sahara had 1.5 billion acres of tropical forest. In 2000, 126 million acres had disappeared due to land clearing by farmers and loggers. Côte d'Ivoire and Madagascar lost 90% of their forests. Half the continent's original forests have disappeared.

African countries have started forest services to protect tropical forests. Logging companies are now using scientific tree farming and replanting programs to protect and renew forests.

Endangered Animals

Deforestation destroys animal **habitats**. Hundreds of species in Madagascar exist nowhere else in the world and are threatened by **extinction**. Population growth has pushed farmers into forests to find land. Savannas, home to giant herds of lions, elephants, and giraffes, are being plowed for farmland. Many species are quickly dwindling in numbers.

Hunting greatly decreased Africa's wildlife during the colonial era. Hunters still pursue African game for sport and profit. The elephant population has decreased from 2 million in the 1970s to 600,000 today due to **poaching**. The Cape Mountain zebra, mountain gorilla, and rhinoceros are also at risk.

Ivory Trade

There were between 5 and 10 million elephants in Africa in the 1930s. Hundreds of thousands of them have been slaughtered for ivory, meat, and sport. The price of ivory soared in the 1970s; gun-wielding poachers started illegally killing elephants for their tusks and more than 80,000 elephants were shot and killed each year.

African elephants were put on the world's endangered species list in 1989. In 1997, Botswana, Namibia, and Zimbabwe sold their stored government stockpiles of ivory to Japan against public outcry.

Great Limpopo Transfrontier Park

At 135,000 square miles, the Great Limpopo Transfrontier Park (GLTP) on the borders of Mozambique, Zimbabwe, and South Africa, is the largest of five national peace parks committed to conserving ecosystems and biodiverse species.

Countries formerly at war are now working together to create and maintain peace parks.

Countries have removed fences and barriers to allow wildlife to follow ancient migration routes.

After 50 million years, rhinos are on the verge of extinction. The poaching of rhinos is very profitable. Rhino horns are used in traditional Asian medicine.

Protecting wildlife and the environment can lead to socioeconomic development and regional peace. Local people are being taught how to benefit from wildlife conservation.

If tourism, game farming of rhinos and other animals, and controlled hunting produce jobs and income, economic stability and peace should result.

Challenges for the Future

Democratic reforms are taking place in Ghana, Nigeria, and Liberia. Expansion of private enterprise has had positive results. Very profitable crocodile farming has brought the species back from very low numbers caused by hunting and trapping. Habitat protection and stricter laws against poaching are helping rhinoceros and elephant populations to recover.

Some countries have created huge game reserves to save endangered species. Tanzania's Serengeti National Park, Kenya's Masai Mara, and Ghana's Kakum National Park have helped animals make a comeback.

Ecotourism, or tourism based on concern for the environment, has become big business in Africa. It has brought millions of dollars into African economies. Rural people work in the reserves as trail guides or become involved in development planning.

Protecting tropical forests is becoming more important. In 1999, the leaders of six countries signed an agreement to preserve forests, a sign Africa is moving in a positive direction.

Practice

Insert T or F to indicate whether the statement is true or false, based on what you've read.

1. _____ Africa south of the Sahara has not yet exceeded its carrying capacity.

2. _____ Deforestation is a minor problem in Africa south of the Sahara.

3. _____ The physical environment has little effect on the creation of transport systems.

4. _____ Countries in Africa south of the Sahara are trying to develop their economies through trading relationships.

5. _____ Over 31 million people in Africa south of the Sahara are in serious need of food.

6. _____ The Sahara Desert is now spreading south into Mauritania, Mali, Niger, Chad, and Sudan.

7. _____ Sudan had over two decades of civil war between the Muslim Arab government in the north and non-Muslim rebels in the south before splitting into two independent nations.

8. _____ Somalia has been without a government since 1991.

9. _____ Killing elephants for ivory tusks is now legal.

10. _____ Rhinos are on the verge of extinction.

Answers

1. F
2. F
3. F
4. T
5. T
6. T
7. T
8. T
9. F
10. T

LESSON

18 EAST ASIA—PHYSICAL GEOGRAPHY

LESSON SUMMARY

Covering nearly 30 percent of the Asian continent, East Asia includes the countries Japan, North Korea, South Korea, China, Taiwan, and Mongolia. Prone to earthquakes, volcanic eruptions, and tsunamis, the region has been shaped by tectonic activity and climatic influences.

Aspire to be like Mt. Fuji, with such a broad and solid foundation that the strongest earthquake cannot move you, and so tall that the greatest enterprises of common men seem insignificant from your lofty perspective. With your mind as high as Mt. Fuji you can see all things clearly. And you can see all the forces that shape events; not just the things happening near to you.

—Miayamoto Musashi, Japanese Martial Arts Master (1584–1645)

Land

Much of East Asia stretching from Central and South Asia to the Pacific Ocean is rugged terrain. East Asia's mountains do not reach great heights. Large plains cover northern and western East Asia. Some plains are covered by deserts, while others carry rich soil for extensive farming. The North China Plain gave birth to Chinese civilization and remains one of the world's most densely populated areas.

Multiple tectonic plates meet in East Asia. The region is therefore susceptible to earthquakes, volcanic eruptions, and ocean flooding. The region is home to the world's highest mountains, the Himalayas, and the symbol of Japan, Mt. Fuji.

Landforms

Tectonic plate movements along the Ring of Fire leave parts of East Asia vulnerable to earthquakes, volcanic eruptions, and ocean flooding.

The People's Republic of China takes up 80% of East Asia and Mongolia takes up 13%. Japan, Taiwan, North Korea, and South Korea make up the rest of the region.

Ring of Fire

Several of the Earth's tectonic plates, including the Pacific, Indo-Australian, North American, South American, and Eurasian tectonic plates, meet along an arc of islands east of China. These mountainous islands are part of the Ring of Fire, the site of frequent volcanic eruptions and earthquakes. Japan has about 50 active volcanoes and hot springs formed by volcanic activity, and experiences 1,000 small earthquakes a year.

DID YOU KNOW?

Mount Fuji was formed by layer upon layer of lava and ash. It has not erupted in 300 years, but could erupt again.

An undersea earthquake generates a tsunami, a huge wave that gets higher as it reaches the coast. Major earthquakes and tsunamis are difficult to predict, so special building methods and emergency preparedness help reduce casualties along the Ring of Fire.

Peninsulas, Islands, and Seas

Many peninsulas and islands dot East Asia. The Korean Peninsula juts southeast from China's Northeast Plain to separate the Sea of Japan, or East Sea, from the Yellow Sea. North Korea and South Korea are mostly mountains surrounded by coastal plains.

Tectonic activity created most of the region's islands and seas. The Sea of Japan was created through subduction millions of years ago. Other bodies of water such as the Yellow Sea, East China Sea, and South China Sea were created through tectonic activity.

Four large, mountainous islands and thousands of smaller ones form the **archipelago**, or island chain, of Japan. Most of these islands were created by volcanic activity over millions of years. Honshu is the largest and central island. Hokkaidō is to the north, while Kyūshū and Shikoku are to the south of Honshu. Most of Japan's major cities are on Honshu, including Tokyo, Kyoto, Kobe, and Hiroshima.

The Sea of Okhotsk is north of Japan. The Sea of Japan and the East China Sea are to the west, the Philippine Sea is to the south, and the Pacific Ocean is to the east and southeast.

DID YOU KNOW?

Japan is an archipelago made up of thousands of islands. It is separated from the Asian continent by the Sea of Japan.

Mountains, Highlands, and Lowlands

Many of Japan's mountains, like Mount Asahi Dake on the northern island of Hokkaidō, are volcanic in origin.

Many mountain ranges spread out from an area of high peaks and deep valleys in western China called the Pamirs. The Kunlun Shan and Tian Shan (*shan* means "mountain") originate in this remote interior region. The Altay Shan further north form a natural barrier between Mongolia and China.

The world's highest mountains to the south and west, the Himalayas, separate China from South Asia. The Himalaya range includes several of the world's highest peaks, including Mount Everest and K2. The Ganges, Yangtze, and other major rivers run through it. There is a wide range of flora and fauna, which varies with the climate and terrain.

> **DID YOU KNOW?**
>
> Mount Everest, the world's highest peak at 29,028 feet, straddles the border between China and Nepal.

The Kunlan Shan becomes the Qinling Shandi as it crosses central China from west to east. The lower Changbai Shan of Manchuria extend into the Korean Peninsula to become the Northern Mountains.

Coastal plains surround the mountain interiors of Japan and Taiwan. Mount Fuji, or Fujiyama, is a dramatic 12,388-foot volcanic cone rising above the plains of Honshu. Mount Fuji is an important spiritual symbol for Japan. Relatively young, it began to rise only 25,000 years ago. Its cone is one of the most perfectly shaped in existence.

The Plateau of Tibet, or Xizang, in China's southwest is East Asia's highest plateau. The plateau averages 15,000 feet in height. Other highlands stretch north and east at lower elevations. The Mongolian Plateau of the far north has extensive highlands and mostly green pasture. The region's only extensive lowlands are China's Northeast Plain and North China Plain.

Broad expanses of wasteland like the deserts and salt marshes of the Tarim Basin lie between Kunlun Shan and Tian Shan. West of the Tarim Basin is the dry, sandy Taklimikan. To the northeast is another desert, the Gobi.

The Gobi Desert in Mongolia and China is an interior lowland plain. It stretches 1,000 miles from southwest to northeast, and nearly 500 miles from north to south. It consists mostly of rock, rather than sand.

> **DID YOU KNOW?**
>
> Nomads in the Gobi build homes of animal skins and blankets stretched on wooden frames that can be taken apart and moved.

Desert storms in the Gobi make life difficult in southern Mongolia and north central China. Less than three inches of rain fall in the Gobi each year.

Water Systems

Two major rivers in China have shaped population patterns and agricultural development. In Japan and Korea, numerous short, swift rivers are sources of hydroelectric power and also create many spectacular waterfalls.

China's Rivers

Landforms and physical processes shaped East Asia's rivers. The region's rivers provide transportation to urban centers, water for drinking, irrigation, hydroelectric power, and rich mineral deposits for fertile soils.

China has tens of thousands of rivers. The major ones start in the Plateau of Tibet and flow east to the Pacific Ocean.

The Huang He, or Yellow, River is northern China's major river system. It deposits almost 60 times as much silt as the Mississippi River and causes extensive flooding. This yellowish brown topsoil, or **loess**, is eroded from western regions by wind and water. Its course has been redirected many times throughout history. It continues to flood and cause many challenges for the region. Loess and river water make the North China Plain a major wheat and soybean farming area.

DID YOU KNOW?

The Hung He is often called "China's Sorrow" because it often floods its banks, killing hundreds of thousands of people.

Central China's Chang Jiang, or Yangtze, River is Asia's longest river at 3,965 miles. It is an unofficial marker between the north and south of China, though it doesn't exactly run through the center of the county. The Chang Jiang flows through awesome gorges and broad plains and empties into the ocean near Shanghai.

The Chang Jiang is a major transport route, providing water for a large agricultural area where more than half of China's rice and grains are grown. The Three Gorges Dam, completed in 2006 and now the largest dam in the world, was constructed with hopes of ending the flooding. It was intended to make more water available for irrigation and extend more hydroelectric power to China's growing population and economy. An engineering and economic success, the dam has simultaneously created new environmental concerns, including the increase of drought and waterborne disease and the triggering of landslides.

The Xi, or West, River is south China's most important river system. The soil deposits of the Xi form a huge, fertile delta. Near the ports of Guangzhou and Macau, this area is one of China's fastest developing areas. The Li River flows through eroded peaks in southern China.

Begun in the 400s B.C., China's Grand Canal is the world's longest artificial waterway.

Over centuries the Grand Canal has been expanded and rebuilt. The Grand Canal moves people and goods along a 1,085-mile course from Beijing to Hangzhou.

Rivers in Japan and Korea

Unlike rivers in China, rivers in Japan and Korea are short and swift. They flow through mountainous terrain, often forming spectacular waterfalls. In the wet season, the rivers in Japan and Korea provide hydroelectric power. Many of Japan's rivers, like the Shinano Tone Rivers, have been altered for irrigation and regulation of water flow. Korea's chief rivers flow from inland mountains west to the Yellow Sea. The Han River flows through South Korea's capital, Seoul. In North Korea, the Yalu, or Amnok, River flows west to form the border with China.

Natural Resources

Mineral resources are abundant in East Asia, but they are unevenly distributed. East Asia's fast-growing economy is proving to be a great strain on the region's energy resources.

China has the most and the widest range of minerals in East Asia—iron ore, tin, tungsten, and gold. There are large oil deposits in the South China Sea and Taklimikan. There are abundant coal deposits in northeast China. Coal is also mined in Mongolia and the Korean Peninsula. North Korea has iron ore and tungsten. South Korea has few minerals, but graphite is found there. Taiwan has small mineral reserves.

Productive farmlands are unevenly distributed across East Asia. About 25% of the land is suitable for farming in South Korea. The country produces two crops per year in the prime farmland of the coastal south. North Korea has increased its cultivated land through irrigation, fertilizers, and mechanized tools.

Mongolia can use only less than 1% of its land for crops. Japan has limited farmland and poor soil. Less than 25% of Taiwan's land is suitable for farming.

Seafood farming has become a major industry in East Asia. Japan, South Korea, Taiwan, and China have the world's biggest deep-sea fishing industries.

DID YOU KNOW?

Environmental damage has resulted from China's rapid economic growth. Air pollution levels in China's larger cities are among the world's highest.

Rice Production

Rice is an important part of the culture and economies of East Asia. It has been grown in the region for over 10,000 years. More than 90% of the world's rice is produced and consumed in Asia. China's "rice bowl" yields up to four harvests per year, making it the world's leading rice producer.

Practice

Fill in the blank with the correct word from the information in the preceding paragraphs.

1. ___________ has the most mineral wealth and energy needs in East Asia.
2. The ___________ is northern China's powerful major river system.
3. Asia's longest river is the ___________.
4. Mount Everest in the ___________ is the world's highest peak, at over 29,000 feet.
5. Mount ___________ is the spiritual symbol of Japan.
6. Japan is a(n) ___________, or island chain.
7. An undersea earthquake generates a(n) ___________, a huge wave that gets higher as it reaches the coast.
8. The Pacific, Philippine, and Eurasian ___________ plates meet along an arc of islands east of China and cause volcanoes and earthquakes.
9. ___________ takes up 80% of East Asia.
10. Less than 1% of ___________ land is suitable for farming.

Climate and Vegetation

Latitude, physical features, and seasonal wind patterns influence East Asia's climates and affect how people live in different landscapes.

Chilling northwesterly winds sweep from Siberia across the Sea of Japan, picking up abundant moisture along the way. In the harshest climates, people and animals on East Asia's last frontier of Hokkaidō find ways to adapt amid howling winter winds.

DID YOU KNOW?

Mongolian nomads survive one of Earth's harshest climates. In most of the country, average temperatures stay near or below freezing year-round.

Climate Regions

East Asia's natural vegetation parallels the region's climate zones, which are influenced by latitude and physical features.

Physical features like mountains, highlands, and coastal areas shape climates, which range from tropical to subarctic. Dry highlands and grasslands dominate the north and west. Humid and temperate forests prevail in the south and east.

Midlatitude Climates

The northeastern part of East Asia, including the northern parts of Korea and Japan, has a cooler, humid continental climate. Summers are warm but winters are long, cold, and snowy.

Southeastern parts of East Asia including Taiwan, parts of China, South Korea, and Japan have a humid subtropical climate with long, hot summers and heavy rains.

The humid continental and humid subtropical climates have needle-leaved and broad-leaved evergreens and deciduous trees. Bamboo grows in warmer areas; it is one of the fastest-growing plants on Earth. It is used for food and herbal medicine, and home, bridge, and skyscraper construction. Bamboo is the only food of the rare giant panda.

DID YOU KNOW?

The mulberry tree, whose leaves provide food for silkworms, the bamboo tree, and the tea bush are important to East Asian culture and economy.

Desert and Steppe Climates

Far from the moist winds of the coast, deserts spread across inland northern China and Mongolia. Moisture is blocked by the rain shadow effect. The Gobi and Taklimikan are often cold and windy. Temperatures can fall 55°F from day to night. Gobi Desert temperatures range from 100°F to –30°F.

Sparse trees and grasses are the natural vegetation of the large steppe climate east of the deserts and in most of Mongolia.

DID YOU KNOW?

Parts of Northwestern China, including the capital city of Beijing, experience seasonal dust storms caused by erosion of the deserts.

Highland Climates

Climate changes with elevation in mountain areas. The higher the elevation, the cooler the temperature. East Asia's highlands are cool or cold. On the Plateau of Tibet with elevations of 13,000 to 15,000 feet, the average annual temperature is 58°F. Small alpine meadows with grasses, flowers, and trees dot the lower mountain slopes. Above the tree line, where no trees grow, only mosses and colorful lichens live.

Tropical Wet Climate

Hainan, off China's southeast coast, has a tropical wet climate. It has high year-round temperatures and very rainy summer **monsoons**. In tropical areas, palms and tropical hardwoods thrive alongside broad-leaved evergreens and tropical fruit trees. Lush rain forest covers most of Hainan.

Seasonal Weather Patterns

Prevailing winds and ocean currents create seasonal weather patterns in East Asia. The air mass above the Asian continent and the one above the Pacific Ocean meet in East Asia. These moving air masses become prevailing winds or monsoons that blow in one direction half the year and then switch directions.

The summer monsoon in East Asia blows from southeast to northwest bringing heat and humidity from the Pacific Ocean. From April through October, intense downpours provide more than 80% of the region's annual rainfall.

From November to March, the winter monsoon brings cold arctic air that blows northwest to southeast. Along the coast, these winds pick up moisture in the Sea of Japan and bring snow to Japan and the Koreas.

East Asian economy relies on summer monsoons to water crops. Late monsoons or less than heavy rain causes crop failure. Too much rain means floods.

The warm **Japan Current** flows north along the coasts of the Japanese islands, adding moisture to the winter monsoon and warming the land.

The cold Kuril Current flows southwest to the Bering Sea along the Pacific coasts of Japan's islands. It brings harsh, cold winters to Hokkaidō's east coast. In summer when the cold ocean current meets the warm one near Hokkaidō, a dense sea fog develops.

Warm, humid air over the tropical ocean produces violent storms called **typhoons**. Typhoons form in the Pacific and blow across coastal East Asia. Parts of Japan experience five or six typhoons per year. Like hurricanes in the western Atlantic and Caribbean, typhoons are usually most severe between late August and October. On occasion, a winter typhoon brings welcome rains during the normal dry season.

Practice

Insert T or F to indicate whether the statement is true or false, based on what you've read.

11. _____ Monsoon winds change direction when they come into contact with warmer or cooler air in their path.

12. _____ Cold, moist winds from Siberia sweep across the Sea of Japan affecting conditions as far as Hokkaidō.

13. _____ Too much monsoon rain can mean drought in East Asia.

14. _____ Sparse trees and grasses are the natural vegetation of the large steppe climate east of the Asian deserts and in most of Mongolia.

15. _____ The humid continental climate of East Asia's northeastern region is characterized by warm summers and long, cold winters.

16. _____ The Plateau of Tibet has a highland climate.

17. _____ East Asia has humid continental and humid subtropical climates.

18. _____ Hokkaidō has a tropical wet climate.

19. _____ Warm humid air over the tropical ocean produces violent storms in the Pacific called typhoons.

20. _____ Humid and temperate forests prevail in the southern and eastern portions of East Asia.

Answers

1. China
2. Huang He, or Yellow, River
3. Chang Jiang, or Yangtze, River
4. Himalayas
5. Fuji
6. Archipelago
7. Tsunami
8. Tectonic
9. China
10. Mongolia
11. F
12. T
13. F
14. T
15. T
16. T
17. T
18. F
19. T
20. T

LESSON 19

EAST ASIA—HUMAN GEOGRAPHY (PART I)

LESSON SUMMARY

Powerful dynasties ruled China since 2000 B.C. The growth of the enormous Chinese empire influenced the cultural development of East Asia. Today, political and economic differences divide the region. China and North Korea are communist, while Japan, South Korea, and Taiwan have capitalist, free market economies.

The most interesting thing about my entire trip to China was there are now 40 million electric scooters and bicycles in China. They all take their little batteries upstairs at night to charge up and bring them back down in the morning and plug them back into their scooters and off they go.

—Jon Wellinghoff, U.S. Federal Energy Regulatory Commission (April 2008)

Countries

China, Japan, Mongolia, North Korea, South Korea, and Taiwan.

China

China's economy and culture are undergoing changes as people migrate from rural to urban areas.

Population Patterns

Cities and rural areas are changing as large numbers of Chinese move to urban areas.

People

About 92% of China's 1.3 billion people descend from the powerful Han family and ethnic group. The Han dynasty and culture, or Middle Kingdom, dominated China from 206 B.C. until A.D. 220. The rest of the population belongs to 55 other ethnic groups.

Taiwan is an island off the southeast coast of China. Taiwan's original indigenous peoples are related to peoples of Southeast Asia and the Pacific. Many Chinese migrated to Taiwan hundreds of years ago.

The ancestors of today's Mongols ruled the world's largest land empire, which stretched from China to Eastern Europe.

Density and Distribution

The North China Plain is one of the most densely populated regions of the world. Over 90% of the people live on one-sixth of the land. Most people inhabit the fertile valleys and plains of the great Huang He, Chang Jiang, and Xi rivers. The urban centers of Beijing, Shanghai, Tianjin, and Guangzhou lie in river valleys or coastal plains.

DID YOU KNOW?

Mongolia has fewer than 4 people per square mile.

Urban Growth

Most Chinese people live on farms. Still, millions of Chinese people migrate to overcrowded urban areas. To curb migration to cities, China has built agricultural towns in remote areas with social services and a better quality of life.

Population growth led the Chinese government in 1979 to limit each family to only one child and population growth slowed. The policy is no longer strictly enforced and population growth is rising again.

History and Government

China has experienced powerful transformations throughout its long history. It has had long-ruling dynasties and been a strong force in East Asia.

Early History

Chinese culture originated 5,000 years ago in the Wei River valley and it spread to surrounding areas. The Shang Dynasty took over the North China Plain in 1600 B.C. and kept China's first written records. As with all successive **dynasties**, the Shang faced rebellions by local lords, attacks by Central Asian nomads, and natural disasters.

The Zhou ruled for about 800 years from about 1045 B.C. The Zhou dynasty began making iron tools and spreading Chinese culture and trade. Confucius, or Kongfuzi, taught a system of thought based on discipline and moral conduct. Laozi, or Lao-Tzu, founded Taoism, a philosophy of living in simplicity and harmony with Nature.

In the A.D. 200s, the Qin Shi Huang Di united all of China and started building the Great Wall to stop attacks from Central Asia. In the Han and Tang dynasties, traders and missionaries spread Chinese culture throughout East Asia.

DID YOU KNOW?

Reaching approximately 5,500 miles from Shanhaiguan in the east to Lop Nur in the west, the Great Wall consists of several walls built throughout China's early history. Most of what remains today was constructed during the Ming Dynasty.

The Qing ruled China from the mid-1600s to the early 1900s. By the 1600s, Western countries attempted to set up shipping routes to East Asia; the Chinese rejected foreign efforts to penetrate their rich silk and tea markets. In the 1800s, Western warships were used to force China to open ports. By the 1890s, European and Japanese governments each claimed spheres of influence or large areas of land where only that country retained exclusive trading rights.

DID YOU KNOW?

Silk trading began in China as far back as the Han Dynasty (206 B.C.–A.D. 220). The Silk Road refers to 4,000 miles of trade routes used by traders and merchants in Asia for almost 3,000 years.

Revolutionary China

In the 1900s, East Asia was involved in two world wars. China had its own internal conflicts.

In 1911, Sun Yat-sen, or Sun Yixian, ended the rule of Chinese emperors. Chiang Kai-shek, or Jiang Jieshi, formed the Nationalist government of the Republic of China. Chiang's Communist rival Mao Zedong gained the support of Chinese farmers. The Communists gained power in 1949 after years of civil war. Mao and the Communists set up the People's Republic of China on the mainland. The nationalists fled to Taiwan and continued the Republic of China.

Economic Changes

In Mao's Great Leap Forward campaign of the 1950s, large government-owned farms replaced small cooperatives. Not enough food was produced, millions starved, and the economy fell apart.

In the 1970s, Deng Xiaoping and other communist leaders permitted private ownership of businesses and farms. China began to modernize and welcomed foreign business and technology. The Communist Party welcomed free enterprise zones to expand capitalism, but not political freedom. Meanwhile, Taiwan has built a powerful export economy and expanded democratic reforms.

A Tale of Two Chinas

Taiwan and China have wanted to unify since the 1950s, but on their own terms. By the 1990s, Taiwan was an economic force. Today, Taiwan has invested billions of dollars in the factories of mainland China. China and the world depend on Taiwan for computer and electronic parts.

Free Mongolia

Mongolia was loyal to China's Qing dynasty until it fell in the revolution of 1911. Then, Mongolia felt no obligation to the new China. Mongolia was a Soviet-influenced Communist state from 1924 to 1991. When the Soviet Union and communism fell, Mongolians adopted a democratic constitution that produced free elections and economic reform.

Culture

Chinese culture is a mix of modern and traditional practices. Its influences include Confucianism, Buddhism, and communism.

Education

Traditionally, the wealthiest Chinese learned to read and write. In an effort to end academic elitism and narrow the cultural gap, requirements for education were increased under Communist rule.

DID YOU KNOW?

In the chaos of the Chinese Communist Cultural Revolution of the 1960s, many historic and cultural sites were destroyed and people thought to be enemies of Mao Zedong were persecuted or executed.

Language and Religion

Most people in China speak the Mandarin dialect of the Han Chinese language. The Chinese government discourages religious practices. The traditional Chinese New Year is celebrated, however, for over a week. Many Chinese people say they are **atheists**. Others hold on to their traditional Buddhist, Confucian, and Taoist faiths. The Chinese government restricts religious practices of the Buddhist population of Tibet.

DID YOU KNOW?

Tibetans can be arrested for owning photos of Tibet's exiled Buddhist spiritual leader, the Dalai Lama.

Arts

Ancient Chinese pottery described human relationships and the beauty of nature. Traditional Chinese opera uses elaborate costumes, music, and acrobatics or martial arts. In the Tang dynasty, potters made fine, thin porcelain we now call china.

Japan

Japan's mix of Asian and Western cultures has helped make the country globally influential. Ancient moats and palaces sit next to modern glass skyscrapers in the Japanese capital of Tokyo.

Population Patterns

Japan's ethnically homogeneous and highly urban society shapes population patterns.

Most of Japan's people live in cities on the relatively lowland seacoasts and in valleys and plains.

People

The population of Japan is ethnically **homogeneous.** About 99% of ethnic Japanese are descendants of Asian immigrants who centuries ago crossed the Korean Peninsula to reach Japan. The aboriginal Ainu were forced north; some still live on Hokkaidō, Japan's northernmost and second largest island.

Density and Distribution

Japan has limited land and the population density is about 875 people per square mile. The central part of the country consists of forest and mountains, so only the valleys and coastal plains were settled. About 78% of Japanese people live in coastal urban areas like the Tōkaidō corridor, a series of cities crowded along the Pacific coast of Honshu between Tokyo and Kōbe.

DID YOU KNOW?

Tokyo is the world's most populous urban area, with more than 35 million people.

Urbanization has changed Japan. Japan's busy city streets have towering skyscrapers and giant electronic advertising signs. City apartments are small. Because the population is so dense and land is so limited, suburban homes in Japan are relatively small, but many have carefully landscaped gardens.

The urban areas of Tokyo have steadily expanded in the twentieth century into outlying areas.

Japan has adapted to crowded conditions by building expressways and train systems. Great for long distance travel, the Shinkansen bullet train reaches over 160 mph on the Tōkaidō corridor.

History and Government

Japan has been shaped by isolation from and interaction with other cultures throughout its history. During this last century, Japan has maintained cultural traditions and modernized.

Early History

In the A.D. 400s, the Yamato dynasty united the ruling families and adopted Chinese philosophy, writing, art, sciences, and governmental systems. Korean scholars also influenced Japan.

In the 1100s, local nobles started fighting for control of Japan. Minamoto became Japan's first **shogun**, or military ruler.

DID YOU KNOW?

Shoguns were supported by professional warriors called **samurai**. Emperors officially ruled Japan, but samurai helped powerful shoguns govern until the late 1800s.

Concerned about European invasion, Japanese shoguns sharply restricted foreigners. Ultimately, only a few merchants were allowed in Nagasaki. Japan remained isolated for the next 200 years.

In 1854, using "gunboat diplomacy," Commodore Perry and the United States Navy forced Japan to trade with the United States. Rebel samurai forced the shoguns to return power to the emperor.

DID YOU KNOW?

During this Meiji Restoration, Japan modernized its government, economy, military forces, education, and legal systems along Western lines.

Modern Japan

From 1890 to 1940, Japan went from a feudal country to a modern one. Education improved and the economy grew quickly.

Japan fought China in 1894 and 1895, winning islands like Taiwan, then called Formosa. Japan fought Russia in 1904 and 1905, winning control of Korea and rights to Manchuria and the large Russian island of Sakhalin. Japan's quickly growing industrial base and navy began to concern European powers.

As Japan joined the Allies in World War I, its economy reached record prosperity. As Japan steered toward democracy, military leaders took control of the Japanese government. They invaded Manchuria in 1931 and China in 1937 and signed a pact with Nazi Germany in 1936.

Japanese leaders believed they would create a "New Order" in the Pacific by expelling Western empires. After four years of fighting in World War II to dominate the Pacific, Japan lost more than 200,000 lives and suffered horrific damage from two atomic bombs dropped on the industrial centers of Hiroshima and Nagasaki.

Within months of the bombings, Japan surrendered its military and territories at the end of World War II in 1945. Japan became a democracy and rebuilt its industrial base with American investment and technical expertise. By the 1970s, Japan was a global economic power with worldwide links to business and trade.

DID YOU KNOW?

Japan is the third largest economy in the world and it sells high-value products to almost every country in the world.

Education and Healthcare

Education is highly valued in Japan; it is compulsory to the age of 15. From a very early age, students focus on getting into good schools. High school curricula focus on preparing students for university.

Improved healthcare has brought life expectancy in Japan to 82. However, the aging population strains the healthcare and social service systems. Government provides healthcare, including preventive services, for those who do not get such coverage through their employers.

Language and Religion

The Japanese language developed in isolation, but may have some Korean and Mongolian influences. Many Japanese words and some of its writing system came from China. Western languages including English have influenced Japanese.

Although Japanese people incorporate Shinto, a traditional Japanese religion stressing reverence for nature, and Buddhist traditions into their holiday and religious ceremonies, most Japanese people claim no religious affiliation.

Arts

The tea ceremony, formal landscaping, and Japanese kabuki theater are all Japanese art forms. Poetry such as **haiku** was popular among the educated classes of ancient Japan. Noblewoman Lady Murasaki Skikibu wrote one of the world's first novels in A.D. 1010. Her *Tale of Genji* told of the life and loves of a prince at the emperor's court.

Family Life and Leisure

The family is regarded in Japan as a source of stability and strength. Ancient beliefs—for example, that being part of a group is more important than individuality—remain strong. Families are smaller now in Japan as the birthrate is declining.

North Korea and South Korea

North Korea and South Korea share similar histories but have been moving in very different directions politically and economically.

Population Patterns

Physical geography and an ethnically homogeneous population affect population patterns in North Korea and South Korea. North Koreans and South Koreans have been migrating to other countries.

People

Korea traces its origins to the early peoples of Northern China and Central Asia. Koreans have experienced long periods of foreign rule and the division of the peninsula into communist North Korea and democratic South Korea. There are no indigenous minority peoples living in North Korea. Small groups of Chinese and Japanese people have lived in South Korea since the mid-1900s.

Density and Distribution

Most people in North Korea and South Korea live on the coastal plains that wrap around the peninsula's mountainous interior. About two-thirds of the Korean population live in growing cities like Seoul, South Korea, and P'yŏngyang, North Korea.

The average population density of South Korea is 1,260 people per square mile. North Korea averages much less at 518 people per square mile. In North Korea, higher population densities occur on coastlines.

North Korea was controlled by the Soviet Union after World War II. The communists focused on rapid industrialization. People moved to cities for jobs and there was a farm labor shortage. After the Korean War (1950–1953), more people moved to cities.

DID YOU KNOW?

North Korea is now about 60% urban. People in South Korea have moved to coastal cities to look for work and South Korea is now about 80% urban.

In the mid-1900s, many people fled communist North Korea for South Korea and other countries. Many Koreans emigrated to Canada and the United States seeking political and economic freedom. South Korea's population doubled between 1950 and 1990. South Korea today has 48 million people, about twice as many as North Korea, where there is a much lower standard of living.

At the end of the Korean War, North Korea and South Korea were divided by a demilitarized zone (DMZ). No official peace treaty has ever been signed and both sides have a large military presence along the DMZ, where soldiers can see each other through binoculars. The two sides have begun to cooperate on trade and tourism despite tension.

History and Government

China and Japan have been aggressive neighbors to Korea. The recent division of North Korea and South Korea remains an important factor in the development of the two countries. North Korea today experiences harsh rule.

Early History

Around 1200 B.C., Chinese settlers brought their culture to neighboring Korea. Buddhism later spread from China to Korea and became Korea's major religion. Over the centuries to follow, the Silla and Koryo dynasties united the Korean Peninsula.

Neighboring countries have invaded and fought over Korea. Mongolia occupied Korea in the 1200s and 1300s. Around 1300, the Chinese seized control of Korea and introduced Confucianism as the model for Korea's government, education, and family life. Japan repeatedly invaded Korea during the 1500s, the height of Korean civilization.

Korea remained independent until the late 1800s, when other countries became interested in it. China and Japan fought over Korea in the Sino-Japanese War of 1894–1895. Russia and Japan fought for control of Korea in the Russo-Japanese War of 1904–1905. Japan defeated China and annexed Korea to its expanding empire in 1910.

Japan harshly administered over Korea and tried to replace the Korean language and culture with Japanese language and culture. Japan occupied Korea until the end of World War II in 1945.

Divided Korea

After World War II, Korea was divided into U.S.-backed South Korea and Soviet communist-ruled North Korea. North Korea invaded South Korea in 1950, in an attempt to unite the country. By June 1951, both sides dug in at the 38th Parallel. The stalemate ended with a truce in 1953. Millions of Koreans had died and both countries were devastated. North Korea and South Korea remain separated.

Following World War II, North and South Korea experienced harsh autocratic rule. In the early 1960s, Major General Park Chung Hee took power in South Korea. He was assassinated in 1979. South Korea's economy grew, but political freedoms were limited under his regime. South Koreans often protested his rule and there were violent confrontations between the government and the pro-democracy movement.

North Korea's centralized government is controlled by the Korean Worker's Party (KWP). All government officials are members of this party; it is the only official party in North Korea. Very little is known about the North Korean government today. Kim Jong Il controlled the government and country from 1994 to December 2011. His son, Kim Jong Un, was announced as the new leader of North Korea. His rule will continue the Kim-family "cult of personality" in North Korea.

Talks between North Korea and South Korea commenced in 2000, but broke off two years later when North Korea reactivated its nuclear reactor.

Strict government rule has caused acute food shortages and mismanagement of essential resources in North Korea.

DID YOU KNOW?

About 2 million North Koreans have died from food shortages since the mid-1990s.

There are more than 200,000 political prisoners in North Korea and reports of human rights abuses in slave labor camps.

Culture

Ancient ties have created a shared culture, while modern political divisions have fueled differences in North Korea and South Korea.

Education has improved in South Korea since World War II. Most children attend middle and high school and university enrollment is increasing. The primary role of education in North Korea is to teach communist ideology. South Korea rebuilt its healthcare system with the assistance of the United Nations following World War II. Life expectancy is now 77 years in South Korea.

DID YOU KNOW?

North Korea provides healthcare to its citizens, but the country still has inadequate food, water, and heating supplies.

Korean is the language spoken in North Korea and South Korea. The language is vaguely related to Japanese and has some borrowed Chinese words.

The Korean way of life is based mainly on Confucianism. People also practice Buddhism, Christianity, and Cheondogyo, a combination of these religions.

Both countries have similar traditional arts. Artists made graceful vases with a pale green glaze called celadon during the Koryo dynasty, which today are still highly valued throughout the world. Buddhist temples contain many statues and sculptures in stone, bronze, or jade. Local woods and granite are used to build Buddhist temples.

The arts are influenced by the different political atmospheres of North Korea and South Korea. Communist ideology shapes culture and art in North Korea. South Korean art has been shaped by many elements of Western culture.

Practice

Fill in the blank with the correct word from the information in the preceding paragraphs.

1. ___________ temples in Korea often contain stone, bronze, jade, wood, or granite.
2. Talks between North Korea and South Korea ended in 2002 when ___________ Korea reactivated its nuclear reactor.
3. Since the end of World War II, ___________ Korea has been influenced by the United States.
4. Kabuki is a traditional form of ___________ theater.
5. ___________ has the third largest economy in the world.
6. ___________ is an island off the southeast coast of China with a powerful export economy.
7. ___________ ruled the Communist People's Republic of China from 1949 until 1976.
8. Emperors officially ruled Japan, but the warrior class of ___________ helped powerful shoguns govern until the late 1800s.
9. The Chinese philosophy of ___________ teaches discipline and obedience.
10. Since the late 1970s, China has created ___________ zones to expand capitalism.

Answers

1. Buddhist
2. North
3. South
4. Japanese
5. Japan
6. Taiwan [formerly Formosa]
7. Mao Zedong
8. Samurai
9. Confucianism
10. Free enterprise zones

LESSON

20 EAST ASIA—HUMAN GEOGRAPHY (PART II)

LESSON SUMMARY

East Asia's increasing participation in the global community and diffusion of the region's cultures still has profound impact in the world. East Asia's growing and aging populations bring future challenges.

East Asia has prospered since the end of the Vietnam War, and Northeast Asia has prospered since the end of the Korean War in a way that seems unimaginable when you think of the history of the first half of the century.

—William Kirby

East Asia and the United States

Migration and trade continues to bring East Asia and the United States together. Over 60% of high technology imports to the United States come from East Asia. Cars, motorcycles, computers, MP3 players, and other high-tech devices are often produced in East Asia and sold in the United States and throughout the world.

Chinatowns

The first major Chinese immigration boom in the United States started in the 1840s. Many immigrants worked in gold mines and railroad construction. Asians often faced discrimination due to cultural and language differences. Chinese people often settled together in Chinatowns. Many of these Chinatowns were located on the west coast, especially California. There are about 25 Chinatowns in the United States today.

Economy

The countries of East Asia are experiencing rapid economic change as they emerge from the Asian financial crisis of the 1990s and adjust to a global economy.

China

China is East Asia's most rural economy. About 40% of China's workers are farmers. China is a leading producer of rice, wheat, tea, soybeans, cotton, and silk.

Since the establishment of the Communist People's Republic of China in 1949, agriculture has changed. Large farming **communes** were established in 1958 during communist Mao Zedong's Great Leap Forward. Members shared equally in work and products. The government decided what and how much to grow. Crop production plummeted and the result was a national famine. In the 1980s, the Chinese government encouraged small household-run farms. Farmers could profit from extra crops or animals. Currently, growth in the industry and service sectors is rapidly outpacing growth in agriculture as many people move from rural areas to major cities in search of a better life.

Mongolia

Most of the land in Mongolia is used for livestock grazing. Privatization of some of the government-owned farmland has led to economic growth. Harsh winters and drought have made the transition to a market economy difficult.

South Korea and North Korea

Only 7% of South Koreans are farmers. They often work on small family farms. Because of migration to urban areas, there is a farm labor shortage. To compensate, modern machinery and efficient farming methods are now being used.

About 36% of North Koreans work in agriculture. Household farms are organized into cooperatives, with the communist government controlling crop production and the rationing of agricultural products. North Korea cannot meet its own needs for rice, the country's major crop.

Japan and Taiwan

Physical geography is a challenge for farmers in Taiwan and Japan. Because Japan is mountainous, Japan uses terracing, modern machinery, fertilizers, and irrigation to raise crop yields. The government provides farmers with financial support so they have as much income as urban dwellers. Even with technology, Japan does not have enough farmland to support its population. Taiwan is mountainous, too, so terracing is often used on farmland to grow rice, sugarcane, tea, bananas, and pineapple.

Industry

Industrial growth in East Asia has been affected by political events, government policies, and changes in the global economy.

Japan

Japan developed its industries after World War II through advanced technology and an educated workforce. Japan was soon a leading producer of cars, computers, telecommunications, electronic equipment, and other consumer products. Its high-quality goods made Japan a global economic power.

Japanese banks failed when they could not collect on loans to risky businesses. Domestic production dropped, the stock market and real estate values plummeted, and unemployment soared. Implementing reforms from 2001 to 2005, Japan recovered and was back to a solid 4% annual growth in real **gross domestic product (GDP)**.

South Korea and North Korea

South Korea moved from an agricultural to an industrial economy after the Korean War. By the 1980s, South Korea was exporting steel, ships, electronics, and motor vehicles. Financial reforms and international aid helped South Korea build its economy after the slump in the 1990s.

In North Korea, government-owned heavy industries produce machinery, military equipment, and chemicals. **Consumer goods** suffer because so many resources go into military production.

DID YOU KNOW?

When economic aid fell with the Soviet Union, North Korea's industrial output fell by 50%.

North Korea's communist leaders were forced to trade with other countries.

North Korea and South Korea agreed to trade in 2000. Relations chilled when North Korea reactivated its nuclear weapons in 2002. In 2006, North Korea fired nuclear missiles into the Sea of Japan despite warnings by the United States and other countries that economic sanctions would follow.

Taiwan

Taiwan is one of the world's leading export-based economies. Profits from agriculture have been invested in the manufacture and export of electronics, plastics, and textiles. Taiwan's economic boom made it a major trading country. About 67% of people in Taiwan work in service industries. Technology-based products are replacing traditional manufactured products.

China

The Chinese government controls the major textile, clothing, footwear, toy, and plastics industries. State-run factories lack updated technology and performance incentives. China has made reform and **privatization** of unprofitable state-run industries a top priority.

China has undertaken some market reforms; privately run small businesses, foreign companies, and foreign investment are now welcome.

China's economy and standard of living have grown sharply due to market reforms. There is still a large economic gap separating industrial areas on the coast and poor agricultural areas in the interior. People are moving to urban areas for work, hoping China's economic expansion will continue.

The Chinese territories of Hong Kong and Macau are major industrial and trading areas. Hong Kong provides great wealth for China and Macau has a profitable market economy that also benefits China.

Transportation and Communications

Japan, South Korea, and Taiwan have nationwide highway and railway networks. The rest of the region is not as developed in terms of transportation networks.

Improvements in transportation and communications networks will link the countries of East Asia and contribute to economic growth.

The China-Tibet Railway, built in 2006, will, according to China, boost Tibet's economy by bringing jobs, trade, and tourism. Conversely, more Chinese people will be drawn into Tibet. That, critics say, will dilute Tibetan culture.

China's rivers provide routes from inland areas to seaports. Shanghai lies near the mouth of the Chang Jiang River. Large ocean-going ships travel inland to Wuhan in Central China. Other important ports are Tianjin on a tributary of the Huang He River and Guangzhou on the Xi River.

Seaports and merchant marine fleets used for commercial transport are vital to export trade. The ports of Hong Kong, China; Chiba, Japan; and Nagoya, Japan, are extremely busy.

In North Korea and China, communist governments control communications, the media, and access to the Internet. People in democratic Japan, South Korea, and Taiwan enjoy a free press and most own radios, televisions, telephones, and cell phones.

Trade and Interdependence

East Asian countries have become increasingly interdependent, but trade disputes and political differences affect these relationships. China, Japan, South Korea, and Taiwan are members of the **Asia-Pacific Economic Cooperation** (APEC), which makes sure trade among member countries is efficient and fair. In 2004, China signed a trade agreement with the 10-nation **Association of Southeast Asian Nations** (ASEAN) to create what became the world's largest free trade area in 2010.

Trade Surpluses: Japan

Japan imports raw materials like iron ore and fuels because it has few mineral resources. The government places taxes on many imported finished goods to protect domestic producers against foreign competition.

High global demand for Japanese goods and high import taxes have given Japan a **trade surplus**. Japan's exports make more money than the value of the country's imports. China and Japan are numbers one and two in trade surpluses with the United States.

Trade surpluses bring increased wealth to Japan and lower profits for Japan's trading partners.

The United States and other countries with **trade deficits** have tried to persuade Japan to open its markets.

In 2001, Japan and the United States established the U.S.-Japan Economic Partnership for Growth which supports opening of markets and cooperation on global economic and trade issues. Trade policy complicates Japan's relations with other countries.

Trade and Human Rights: China

Seeking to modernize, China has pursued increased trade with the United States and other market economies. China's undervalued currency keeps export prices low, so there is a flood of cheap Chinese goods being exported to the United States. China's 2006 record trade surplus renewed concerns that Chinese currency values are artificially kept low.

China's recent economic and trade successes have not resulted in human rights reforms. **Dissidents** or citizens who publicly contest government policies, are harshly treated. In 1989, the world saw on television the Chinese government's brutal suppression of democratic student demonstrators in Tiananmen Square, Beijing.

The United States, Japan, and other important trading partners have placed economic or trade restrictions on China for human rights violations. China released some dissidents and the United States lifted sanctions.

In 2000, the United States granted full trading privileges to China. The following year China was admitted to the **World Trade Organization** (WTO), which oversees international trade agreements and settles trade disputes between countries.

People and Environment

Throughout East Asia, rapid industrialization and the burning of fossil fuels has led to severe pollution of the air, land, and water. Most of the land along the river plains has been completely transformed. Rapid industrial and urban expansion and higher standards of living in East Asia are threatening the environment.

Managing Resources

The increased burning of fossil fuels has led to severe pollution and the search for cleaner sources of power. Industrialization and urbanization mean increased demand for electric power. Higher standards of living mean people buy and use more appliances and electronic devices.

Fossil Fuels

East Asia's primary energy source is fossil fuels. China, North Korea, and Mongolia use large coal reserves to create power. Japan and South Korea rely heavily on foreign sources of oil.

Solar panels are widely used in Japan and the government has invested in nuclear power stations. China's massive Three Gorges Dam on the Chang Jiang supplies hydroelectric power to China's interior.

Nuclear Energy

Nuclear power provides about 40% of the energy for Japan, South Korea, and Taiwan. North Korea is thought not to have a single nuclear power facility. China plans to build over 100 nuclear plants. Japan has 54 nuclear reactors, South Korea has 19, and Taiwan 6 nuclear reactors.

People in East Asia are concerned about nuclear radiation. There were nuclear accidents in Japan and South Korea in the 1990s. After a 1999 nuclear accident, Japan completed alternative wind- and solar-powered electricity generators. Some people are concerned tectonic shifts could cause nuclear reactors to crack and release radiation. The March 2011 tsunami struck a nuclear facility in Fukushima, Japan, and caused a complete shutdown and radiation leaks.

Human Impact

The impact of industrial and economic growth in East Asia includes environmental damage, depletion of natural resources, and health risks. China must meet environmental challenges as it grows its industries and economy.

China

Outdated industrial and transportation technology has caused major air pollution in cities. China heavily relies on large supplies of cheap coal that mixes with air-blown dust to cause air pollution and lung disease.

Acid Rain

Acid rain from burning coal is a major problem for China and its neighbors. Japanese forests receive pollution from China. Some of it reaches the Pacific Ocean and parts of the Western United States.

DID YOU KNOW?

Air pollution from China travels across the Pacific Ocean and contributes to the smog over Los Angeles.

Human and Industrial Waste

Rapidly urbanizing countries like China have trouble disposing of waste. Two-thirds of China's cities do not have clean, fresh water. Hundreds of millions of people must boil their drinking water.

Industrial waste is linked to certain types of cancer. In 2000, China for the first time shut down a government-owned factory for environmental reasons—the metal factory in Shenyang was emitting toxic amounts of sulfur dioxide and other chemicals into the air, harming the health of nearby residents.

Deforestation and Desertification

China plants and cuts thousands of acres of forests to meet its demand for lumber. Clear cutting timber leads to deforestation and soil erosion. Trees slow rain runoff. Without trees, flooding occurs. In the late 1990s, heavy rains caused the Chang Jiang and Hung He Rivers to flood, altering the landscape, destroying property, and killing thousands of people.

China has begun planting trees on millions of acres along deforested riverbanks.

Koreas and Taiwan

Negligent industrial controls have led to air and water pollution in North Korean, South Korean, and Taiwanese cities. Untreated sewage contaminates water supplies and threatens the health of humans and wildlife.

Japan Leads the Cleanup

Japan's environmental laws are among the world's strictest. Japan has urged other countries to reduce emissions of carbon dioxide and chlorofluorocarbons (CFCs), which destroy Earth's protective ozone layer.

In 2002, Japan and other industrialized nations signed the Kyoto Treaty and committed themselves to reducing carbon dioxide emissions. Japan's carbon dioxide emissions still increased by 8% over 1990 levels.

East Asia's coastal waters, including the Inland Sea, have been overfished. Many commercial fishing companies are now fishing further off the coast in international waters. Giant factory ships called supertrawlers follow fishing fleets and quickly clean and freeze large catches of fish. Harvesting large catches leads to overfishing, so **aquaculture**, or cultivating fish and other seafood, is emphasized.

Japan is internationally criticized for its whaling practices. Overhunting for whale meat has caused the whale population to seriously decline. Despite a 1986 international treaty limiting whale hunting, Japanese fleets still steadfastly hunt them.

Future Challenges

Due to its location and physical geography, East Asia faces challenges from natural disasters like floods and earthquakes.

China's Huang He and Chang Jiang can flood disastrously. Flood control precautions include building networks of drainage and irrigation channels to transport or redirect water, and constructing dikes, levees, and dams. More than 30,000 of the dams built in China in the 1950s and 1960s are defective and at risk of failure. Severe flooding continues.

Construction of the Three Gorges Dam forced the relocation of two million people, put farms, villages, and ancient temples under water, and destroyed ecosystems. As the water rises, soil and chemical pollutants in abandoned rivers may leach into the river.

East Asian countries will continue to face earthquakes like the ones that hit Taiwan in 1999 and China's Yunan Province in early 2000 and mid-2003. Japan has over 1,500 small earthquakes a year. Underwater earthquakes or volcanoes can trigger tsunamis. In March 2011, a 9.0 magnitude earthquake triggered a tsunami that hit the northeast coast of Japan, killing 25,000 people and causing billions of dollars in damage.

Practice

Insert T or F to indicate whether the statement is true or false, based on what you've read.

1. ____ Japan, South Korea, and Taiwan have command economies.

2. ____ China's Great Leap Forward succeeded in crop production.

3. ____ Japan and North Korea are agriculturally self-supporting.

4. ____ When Soviet aid to North Korea stopped, North Korea's industrial production fell.

5. ____ Migration to urban areas in China is increasing.

6. ____ Japan, South Korea, and Taiwan have nationwide highway and railway networks.

7. _____ In North Korea and China, communist governments control communications, the media, and access to the Internet.

8. _____ The Kyoto Treaty, signed by Japan in 2002, marks a committment to reducing carbon dioxide emissions.

9. _____ Mongolia's fertile soil has made it a leading world food producer.

10. _____ More than half of China's cities do not have clean fresh water.

Answers

1. F
2. F
3. F
4. T
5. T
6. T
7. T
8. T
9. F
10. T

LESSON

21 SOUTHEAST ASIA—PHYSICAL GEOGRAPHY

LESSON SUMMARY

Southeast Asia is made up of the Asian continent south of China and many islands north of Australia. Mainland Southeast Asia stands between South Asia and East Asia. The islands of Southeast Asia, or Maritime Southeast Asia, curve in an arch from southwest to southeast of the mainland. Most of the region's islands belong to the countries of Indonesia or the Philippines. These islands were formed by the collision of Earth's tectonic plates, so they have many active volcanoes.

To travel in Europe is to assume a foreseen inheritance; in Islam, to inspect that of a close and familiar cousin. But to travel in farther Asia is to discover a novelty previously unsuspected and unimaginable.

—Lord Byron

Landforms

Millions of years ago the Eurasian, Philippine, and Indo-Australian tectonic plates collided and formed parallel mountain ranges and plateaus called cordilleras. Tectonic activity including volcanoes created a group of islands called archipelagoes.

Peninsulas and Islands

Southeast Asia runs 1,735,448 square miles along the equator and from the Asian mainland to Australia. The southeast Asian mainland includes the Indochina Peninsula and the Malay Peninsula.

The Indochina Peninsula includes all of Vietnam, Laos, Cambodia, and Myanmar, and part of Thailand. The peninsula is surrounded by the South China Sea, the Gulf of Thailand, and Andaman Sea.

DID YOU KNOW?

Southeast Asia's major rivers are located on the Indochina Peninsula.

The northern portion of Indochina has a humid subtropical climate. The southern part of Indochina has tropical wet and tropical dry climates.

The Malay Peninsula includes parts of Thailand and Malaysia. The peninsula is surrounded by the Gulf of Thailand and the Strait of Malacca and has a tropical wet climate.

The Malay Archipelago includes Indonesia, the Philippines, Singapore, and most of Papua New Guinea. With more than 25,000 islands, it is the largest archipelago in the world. It is made up of several smaller archipelagos, including the Philippines.

Mainland Southeast Asia

About half of Southeast Asia lies on the mainland; the rest is islands.

DID YOU KNOW?

Vietnam, Laos, Cambodia, and Myanmar (formerly Burma) lie completely on the Indochina Peninsula.

Most of Thailand lies on the Indochina Peninsula, but some of it trails south to the Malay Peninsula. Malaysia is a mainland and island country. Malaysia shares the Malay Peninsula with Thailand. The rest of Malaysia is located on the island of Borneo and many other islands.

Maritime Southeast Asia

The **insular**, or island, countries of Southeast Asia are a series of archipelagoes, or island chains, extending from the Indian Ocean to the Pacific Ocean. They include Brunei, East Timor, Indonesia, Singapore, and the Philippines. Many of the islands are not inhabited and do not have names. Most of them have a tropical wet climate.

Brunei, on the northern coast of Borneo, is surrounded almost completely by Malaysia. Spanning some 2,200 miles, Brunei has two distinct, nonadjacent regions: the mountainous west and the lowland east.

Indonesia is the largest island country in the region with 17,500 islands, of which approximately 6,000 are permanently settled. Temperatures in Indonesia remain high year-round, with a wet season between October and April.

DID YOU KNOW?

East Timor was once part of Indonesia, but is now independent.

Singapore is made up of one large island and 62 smaller islands off the southern tip of the Malay Peninsula. The large island is roughly 270 square miles in area, and a mere 50 feet above sea level. Only one degree north of the equator, Singapore's climate is hot and humid with large amounts of rainfall all year.

With a total land area of more than 185,000 square miles, the Philippines are made up of some 7,000 islands. Over 95% of the people in the Philippines reside on 11 of its islands; roughly half reside on the island of Luzon. The Philippines have a tropical marine climate, with a rainy season from May through December.

DID YOU KNOW?

Volcanic islands form when magma flows from faults in Earth where it cools and hardens. The hardened lava slowly builds itself up into a cone. An active volcano eventually builds up enough lava to rise above sea level and become an island.

Mountains and Volcanoes

Mountains and volcanoes dominate the landscape of Southeast Asia and create geographic and political boundaries. The western and northern highlands of Indochina's Peninsula separate the region from India and China.

To the south and east, three cordilleras, or mountain ranges, on the mainland run parallel to each other. These mountain ranges form natural barriers between and within mainland countries. These parallel ranges include the Arakan Yoma in western Myanmar, the Bilauktaung between Myanmar and Thailand, and the Annam Cordillera separating Vietnam from Laos and Cambodia.

Mountains on the islands of Southeast Asia form part of the Ring of Fire, an area of volcanic and earthquake activity surrounding the Pacific Ocean. Some of the mountains in Southeast Asia are volcanoes, and some are active.

More than 300 volcanoes stretch across Indonesia. The Indonesian island of Java is one of the most active areas in the Ring of Fire. This geologic hot spot has 21 of Indonesia's 129 active volcanoes.

The 1883 eruption of Kratkatau in Indonesia caused massive destruction and loss of life.

Java's Gunung Merapi, or "Fire Mountain," is among the most active volcanoes in the world. When Java's tallest and most active volcano started spewing toxic gas, volcanic ash, and molten lava in 2006, Indonesian authorities put alert procedures and evacuation plans into action. Volcanic activity is still monitored in Java so the population can be warned about a coming eruption.

In August 2010, Mount Sinabung in Indonesia, just 40 miles from Sumatra's main city, Medan, erupted for the first time in 400 years.

TIP

Mineral-rich volcanic ash breaks down and becomes rich fertile soil. The soil on these islands makes many productive agricultural areas.

The June 1991 volcanic eruption of Mount Pinatubo in the Philippines was one of the most violent and destructive volcanic eruptions in the twentieth century. The eruption created a large basinlike depression that reduced the mountain's size from 5,275 feet to 4,875 feet. The eruption caused incineration and destruction miles away. More than 300 people were killed.

Mayon Volcano erupted in 1993 and again in 2000. Mount Agung on the small Indonesian island of Bali reaches 10,308 feet.

Plains and Plateaus

Hollowed-out stones, once used for storage by inhabitants, litter a plateau in Laos. The uplands of this region are often lined by mountains.

Water Systems

Southeast Asia's rivers provide food as well as transportation and communication routes. Sediment deposits make rich, fertile soils.

Mainland rivers originate in the northern highlands of Southeast Asia and southern China. Major rivers are the Irrawaddy River in Myanmar, the Chao Phraya River in Thailand, and the Red River in Vietnam.

The 2,600-mile-long Mekong River begins in China and flows along the border of Thailand and Laos before entering Vietnam to form a large delta. Sediment deposits build up the shoreline of the Mekong Delta by as much as 50 feet a year.

DID YOU KNOW?

The Mekong is the longest of Southeast Asia's five major rivers.

Southeast Asia's island rivers are shorter than those on the mainland and they flow in various directions.

Natural Resources

Southeast Asia's natural resources are just as varied as its physical features. Fossil fuels, minerals, and gems are some of the region's natural resources. The flora and fauna of Southeast Asia are valuable resources and some of the world's most diverse.

DID YOU KNOW?

Some traditional workers in Vietnam use rakes to gather salt that has precipitated from seawater.

Fossil Fuels

Southeast Asia has large supplies of coal, oil, and natural gas. Malaysia, Brunei, Indonesia, and Vietnam export large amounts of oil. Indonesia was a member of the Organization of the Petroleum Exporting Countries (OPEC) until 2008.

DID YOU KNOW?

Oil and natural gas deposits off Borneo's northern coast have made the sultan, or ruler, of Brunei one of the world's wealthiest people.

Ownership disputes have stopped oil and natural gas exploration off the coast of the Spratly Islands in the South China Sea. Coal comes from Vietnam and the Philippines.

Minerals and Gems

Southeast Asia has abundant mineral reserves. Indonesia, Malaysia, Thailand, and Laos are leading tin producers and exporters. Indonesia mines nickel and iron and the Philippines mines copper.

Gems provide income for Southeast Asia. Pearls are harvested in the waters off the Philippines. Sapphires and rubies are found in Myanmar, Thailand, Cambodia, and Vietnam.

Most countries take advantage of nature's wealth. Some countries have untapped, underdeveloped, or mismanaged resources: Myanmar is rich in zinc, jade, rubies, and sapphires, but decades of military rule and government control of key industries in Myanmar have led to mismanagement of natural resources and a **black market**.

Flora and Fauna

Southeast Asia's plant life is exotic and diverse. The amazing *Rafflesia arnoldii*, with a three-foot blossom, is the world's largest flower. Thailand cultivates over 1,000 species of orchids. The flowers are a valuable trade commodity. Workers tap rubber trees in Malaysia and process woods like teak from Myanmar for export.

Southeast Asia's fauna is distinctive and varied. Wildlife sanctuaries and national parks in Southeast Asia have roaming elephants, tigers, rhinoceros, and orangutans. Southeast Asia has animals—like Borneo's bearded pig, Malaysia's lacewing butterfly, and the Komodo dragon—that cannot be found anywhere else in the world.

DID YOU KNOW?

The Komodo dragon, native to Indonesia, is the world's largest lizard.

Some of these animals are endangered, or at risk of extinction. Sanctuaries and national parks have helped protect endangered species in some countries.

Practice

Fill in the blank with the correct word from the information in the preceding paragraphs.

1. The Philippine and Indo-Australian tectonic plates collided and formed parallel mountain ranges and plateaus called ___________.

2. The ___________ Peninsula includes parts of Thailand and Malaysia.

3. Maritime Southeast Asia includes many smaller ___________ extending from the Indian Ocean to the Pacific Ocean.

4. The western and northern highlands of the ___________ Peninsula separate Southeast Asia from India and China.

5. This geologic hot spot of ___________ has 21 of Indonesia's 129 active volcanoes.

6. ___________ deposits make rich, fertile soils in Southeast Asia.

7. The ___________ is the longest of Southeast Asia's five major rivers.

8. Southeast Asia's ample fossil fuel reserves include coal, oil, and natural ___________.

9. Sapphires, rubies, and ___________ are harvested in Southeast Asia.

10. The ___________ dragon of Indonesia is the world's largest lizard.

Climate and Vegetation

Southeast Asia's climates support many diverse ecosystems and natural habitats, some of which have disappeared or are endangered due to urbanization and logging.

Moist summer monsoons blow in from the south, bringing abundant rain. Exotic tropical flowers perfume the air of Southeast Asia.

Tropical Regions

Southeast Asia's extensive tropical climates support diverse ecosystems. Island Southeast Asia mostly has a tropical wet climate. Most parts of the mainland and some parts of the islands have a tropical dry or humid subtropical climate.

Tropical Wet Climate

The islands and coastal areas of Southeast Asia have little temperature variation and mostly wet conditions all year. With an average temperature of 79°F, humidity of 80–90%, and rainfall between 79 and 188 inches, Southeast Asia is hot, humid, and rainy.

DID YOU KNOW?

The summit of Mount Isarog in southern Luzon in the Philippines sometimes receives 468 inches of rain a year.

The tropical wet climate of Southeast Asia supports a diverse ecosystem. The Malaysian rain forest with triple canopies and broadleaf evergreen trees dates back millions of years. Between river valleys and higher elevations are several layers of vegetation. Peat swamp forests dominate river valleys. Sandy coastal soil supports shrubs and mangrove swamp forests cover tidal mudflats. Lowlands with poor or shallow soil support forests of tall trees with leathery evergreen leaves. Some of these trees produce an aromatic organic resin compound.

The tropical wet climate makes parts of Southeast Asia particularly subject to large numbers of storm-induced flash floods. The rivers of mainland Southeast Asia have seasonal flooding each year. A **cyclone** is an area of low atmospheric pressure surrounded by circulating winds extending out 10 to 1,000 miles. A typhoon is a tropical cyclone formed in the Pacific Ocean near the equator. Typhoons, like the one that hit the Philippines in 2006, can have winds of 150 to 180 mph, with rain and high ocean waves.

On December 26, 2004, an earthquake in the Indian Ocean created a tsunami that left 225,000 people dead and millions homeless. Banda Aceh, Indonesia, experienced massive destruction.

Singapore

Singapore has transformed from a mostly dense rain forest surrounded by mangrove trees to a sprawling urban area. Natural habitats and **endemic**, or native, species are gone. Most trees and shrubs in Singapore were imported from places like Central and South America.

DID YOU KNOW?

The few, small portions of Singapore's rain forest that have survived are now part of larger for-profit attractions.

Tropical Dry Climate

A tropical dry (winter) climate sweeps southeast across the Indochina Peninsula and southeastern parts of Indonesia. Wet and dry seasons alternate. The vegetation includes tropical grasslands with scattered trees and a few forests.

On the mainland from May to September, summer monsoons bring rain. Winter dry season lasts from November to April. The first few months of this time frame are cool, but the last few months are hot.

DID YOU KNOW?

South of the equator in southern Indonesia, the wet and dry seasons occur in reverse. From May to September, South Pacific trade winds bring the hot dry season.

Monsoons bring rain from November until April. Because winds blow rain to and away from land at different times of the year, the islands frequently have rain on one side part of the year, and on the other side the other part of the year.

Midlatitude Regions

Humid subtropical and highland climates in the midlatitude regions of Southeast Asia support a variety of vegetation. Parts of the mainland of Southeast Asia, including most of Laos, a small portion of Thailand, and northern Myanmar and Vietnam, have a humid subtropical climate. From November to April, cool dry temperatures average 61°F.

DID YOU KNOW?

The elevated Shan Plateau in Myanmar resembles a cool climate and is called "tropical Scotland."

Highland climates predominate in mountainous areas of Myanmar, New Guinea, and Borneo. Cooler temperatures support deciduous forests with moss-covered tree trunks on lower slopes. These **deciduous** trees lose their leaves in autumn. Evergreen **coniferous** forests stand on higher elevations. In the highland climate of Myanmar, there are forests of rhodendrons.

Practice

Insert T or F to indicate whether the statement is true or false, based on what you've read.

11. ____ In Southeast Asia, mangrove swamp forests cover plateaus.

12. ____ Highland climates exist in New Guinea, Borneo, and Myanmar.

13. ____ The Malaysian rain forest has only single tree canopies.

14. ____ Parts of Southeast Asia experience below-freezing temperatures through much of the winter.

15. ____ In the humid subtropical climate of Vietnam, winter temperatures remain above 70°F on average.

16. ____ Rhodendrons, evergreen trees, and mosses can be found in the highlands of Southeast Asia.

17. ____ The world's largest flower has three-foot blossoms and grows in Southeast Asia.

18. ____ Moist summer monsoons blow in from the north bringing abundant rain.

19. ____ The islands of Southeast Asia have a tropical dry climate.

20. ____ The heavily urbanized Singapore we know today was once a land covered with tropical rain forests and mangrove trees.

Answers

1. Cordilleras
2. Malay
3. Archipelagoes
4. Indochina
5. Java
6. Sediment
7. Mekong
8. Gas
9. Pearls
10. Komodo
11. F
12. T
13. F
14. F
15. F
16. T
17. T
18. F
19. F
20. T

LESSON

22

SOUTHEAST ASIA—HUMAN GEOGRAPHY

LESSON SUMMARY

Southeast Asia has a large number of ethnic groups with their own languages and cultures. The region stands at the crossroads of South Asia and East Asia, so it has been influenced by various cultures. Arab traders introduced people in Southeast Asia to Islam. The countries of Southeast Asia are industrializing and urbanizing at different rates. Rich in natural resources, Southeast Asia faces a variety of environmental problems.

I don't have any formula for ousting a dictator or building democracy. All I can suggest is to forget about yourself and just think of your people. It's always the people who make things happen.

—Corazon Aquino, president of the Philippines 1986–1992

Countries

Brunei	Indonesia	Myanmar	Thailand
Cambodia	Laos	Philippines	Vietnam
East Timor	Malaysia	Singapore	

Mainland Southeast Asia

Settlement patterns and regional conflicts have influenced the cultures of mainland Southeast Asia. The people have been influenced by ancient migrations, cultural and political changes, and the blending of traditional and modern lifestyles. Some decades-old conflicts remain, but mainland Southeast Asia retains its timeless beauty and diverse cultures.

Population Patterns

The population patterns of mainland Southeast Asia have been shaped by migrations and conflicts.

People

There have been humans in Southeast Asia for tens of thousands of years. Migrants from western China and eastern Tibet settled the region about 2,500 years ago. On the mainland, the Khmers first settled in Cambodia and Vietnam. The Khmers then migrated into Vietnam, where they now make up about 90% of the population. Some people migrated from China to Vietnam.

The Mons, native to lower Burma, were invaded by the Burma ruler in 1757. The survivors of the massacre fled to Thailand. Today the Mons are integrated into the cultures of Myanmar and Thailand. The Mons integrated into Burmese culture. The Thai people of Thailand descend from the people of southwest China. The Lao people descended from the Thai people who moved into what is today Laos.

Density and Distribution

In Southeast Asia, population densities vary. Laos has 65 people per square mile, while Vietnam has 650 people per square mile. Bangkok, Hanoi, Yangon (Rangoon), Phnom Penh, and Ho Chi Minh City each have populations of more than 1 million people. Population is concentrated in river valleys and coastal plains where there is water, fertile land, transportation, and jobs. People in Southeast Asia are moving to cities to escape political conflict, and to seek jobs and education.

Primate cities serve a country's ports and economic centers, and are often the capital city. The primate city of Bangkok has grown by more than 5 million people since 1991. Since the 1970s, there has been much external migration from mainland Southeast Asia. Between 1975 and 1990, thousands left Vietnam and Laos to flee political oppression and economic crisis. Many skilled and educated workers, needed to sustain economic growth in Southeast Asia, migrated to the United States.

History and Government

Ancient histories and modern political conflicts in mainland Southeast Asia continue to shape the region. European colonial rule led to struggles for independence and democracy that still influence Southeast Asia today.

Early Civilizations

Early peoples of Southeast Asia were highly skilled farmers. Rice was their main staple, as it is today. They were advanced metalworkers.

Southeast Asian cultural traditions, like the worship of ancestors and animal and nature spirits, or animism, developed around the same time. Ancient Southeast Asia had matriarchal societies—power and wealth were passed down through the mother's family.

Funan Kingdom

Traders from India set up trading posts along the Gulf of Thailand during the A.D. 100s. People along the gulf combined Indian traditions with their own and formed the Kingdom of Funan. People in Funan adopted Hinduism and the Indian idea of centralized government under one powerful ruler. Funan was a great maritime power, engaging in trade with India, China, and Persia.

Khmer Empire

Organized agriculture brought wealth to mainland Southeast Asia. In the 1100s and 1200s, the Khmer Empire used a complex system of lakes, canals, and irrigation channels to grow three or four crops of rice a year.

> **DID YOU KNOW?**
>
> Magnificent Khmer architecture like the 800-year-old temple of Angkor Wat was designed to mirror the home of Hindu gods.

Vietnam

The Vietnamese people controlled most of the Indochina Peninsula until 111 B.C. The Chinese then conquered the area and held it until the early A.D. 900s. The Chinese introduced their system of writing and ideas about philosophy and government to Vietnam.

Western Colonization

By the 1500s, Europeans were trading, spreading Christianity, and claiming territory in Southeast Asia. They set up agreed-upon areas of control called spheres of influence. Later, Europeans established colonies in Southeast Asia. The Kingdom of Siam (now Thailand) served as a **buffer state** between rival British and French powers.

> **DID YOU KNOW?**
>
> Siam was the only territory in Southeast Asia to remain free of European rule.

In the early 1900s, the Netherlands, the United Kingdom, France, and the United States replaced small farms with large plantations in Southeast Asia to make huge profits. To meet the need for labor, European owners hired many Indian and Chinese immigrants who settled permanently in Southeast Asia.

Struggles for Freedom

After World War II, Southeast Asians fought for their independence. By 1965, every country in Southeast Asia had won its struggle for freedom from at least one colonial ruler. Some countries suffered from continuing instability. Myanmar's military overthrew its government in the 1980s and instituted harsh military rule known as **martial law** over civilians.

War in Vietnam

In 1954, communist rebels under Ho Chi Minh defeated the French in Vietnam. The Geneva Peace Accord divided the country into communist North Vietnam and non-communist South Vietnam. Hundreds of thousands of people left Vietnam during a grace period after the accord was signed. To stop the spread of communism into South Vietnam, the United States supported South Vietnam beginning in 1962 by committing thousands of armed forces. An estimated 3 million Vietnamese were killed, including up to 2 million civilians. The United States left Vietnam to a unified communist government in 1975.

Cambodia and the Khmer Rouge

Communist forces called the **Khmer Rouge** took over Cambodia in 1975. They forced people out of cities and towns and onto farms. People were not prepared for this change of lifestyle.

> **DID YOU KNOW?**
>
> Over 1 million Cambodians died from overwork, disease, starvation, or execution at the hands of the communist Khmer Rouge.

Culture and Society

Mainland Southeast Asia today reflects the many generations of various cultures in the region. In Vietnam, for example, Chinese, Hmong, Thai, Khmer, Man, and Chan cultural traditions exist alongside Vietnamese culture.

Education and Healthcare

After World War II, literacy and education expanded in mainland Southeast Asia. Laos reorganized its educational system in the mid-1970s and literacy improved. Cambodia still faces a shortage of funds, supplies, and trained teachers.

Mainland Southeast Asia has inadequate healthcare unevenly distributed across the mainland. Most countries and rural areas in the region lack medicine and necessary sanitation. Gastrointestinal diseases, tuberculosis, malaria, and lately, avian flu, have been epidemic. Cambodia, Thailand, and Myanmar have high rates of HIV infection and AIDS.

Language and Religion

Most of the hundreds of languages in the region are part of the Sino-Tibetan or Mon-Khmer families.

Urban residents of Vietnam speak Vietnamese, Chinese dialects, French, or English. The languages of Vietnam, Laos, and Thailand are tonal. Pitch variations help people distinguish between similarly pronounced words. Most people in Myanmar speak Burmese. Over 95% of Cambodians speak Khmer.

East Indian traders brought Hinduism and Buddhism to Southeast Asia, and Arab traders brought Islam. Buddhism is the major religion in Southeast Asia.

Many people with Chinese ancestry follow Confucianism or Taoism. Religions mingle in Southeast Asia. In Vietnam, people blend Buddhism, Confucianism, and in some cases, Catholicism.

Maritime Southeast Asia

Indigenous and outside cultures have influenced the thousands of islands of Southeast Asia. The islands span oceans and seas and have a mix of traditional and modern cultures. As these islands modernize and share in the global economy, they attempt to maintain their cultural traditions.

Population Patterns

Migration and trade have shaped population patterns in Island Southeast Asia. The Southeast Asian islands of Malaysia, Indonesia, Brunei, Singapore, East Timor, and the Philippines have sparsely populated mountainous island interiors and densely populated coastal areas.

People

The islands and peninsulas of Southeast Asia have about 360 million people from various ethnic groups. Many are descendants of people who came from the mainland. The indigenous people of Malaysia, the Malays, first settled Indonesia before arriving on the Malay Peninsula.

Valuable spices attracted traders to the islands of Southeast Asia. In about A.D. 1000, merchants from India brought Hindu and Buddhist religions, art forms, and government that glorified kings. Traders and soldiers from China influenced Brunei, Malaysia, and Singapore, where 77% of the current population has Chinese ancestry. In the A.D. 800s, Arab traders brought Islam; the religion spread widely. In the 1400s and 1500s, European traders influenced Southeast Asia.

Density and Distribution

The people of Southeast Asia live mostly on coastal plains where there are food, jobs, and transportation. Some islands have very high population densities. Java has an amazing 2,375 people per square mile. Singapore, Southeast Asia's smallest country, has an incredible 18,513 people per square mile.

People are migrating from rural areas to the cities of Southeast Asia for better economic and educational opportunities. The Indonesian trend toward urbanization can be seen in its capital of Jakarta; over 10 million people live in this city on the island of Java. To reduce overcrowding, Indonesia's government has relocated millions of people to the country's less densely populated outer islands.

History and Government

The location of Southeast Asia's islands has played an important role in the region's history. Trade has always influenced the development of Southeast Asia's islands.

Early History

From Sumatra, the Srivijaya Empire controlled the seas bordering Southeast Asia from A.D. 600 to 1300. The empire acquired wealth by taxing passing trade ships. Singapore uses the same trade routes today.

Indian and Muslim Arab merchants and missionaries shaped the islands of Southeast Asia. Many people on the islands adopted Islamic ways and converted. Islam quickly spread from the coast to interior areas of the Malay Peninsula and to neighboring islands. In the 1400s, Malacca on the Malay Peninsula was an important seaport and cultural center.

Colonization and Freedom

In the early 1900s, when European countries started colonizing Southeast Asia, the Netherlands claimed most of the islands that today make up Indonesia and called them the Dutch East Indies. The United Kingdom controlled what are now Singapore and Brunei. The United States gained control of the Philippines following a war with Spain in 1898.

DID YOU KNOW?

Over 1 million Filipinos died when Japan invaded the Philippines during World War II.

After World War II, the United States granted the Philippines independence. The Philippines and many other Southeast Asian countries faced years of fighting between communist and conservative forces.

On the islands of Southeast Asia, some ethnic groups have struggled for independence. With assistance from the United Nations, East Timor became fully independent in 2002.

In 1998, Indonesia moved toward democracy after years of dictatorship. The Philippine government, also a democracy, has struggled with corruption, coup attempts, and debt. Singapore follows the structure of a parliamentary republic, but has been ruled by the same party for more than 50 years. Laos and Vietnam are communist states. Brunei, Cambodia, Malaysia, and Thailand are constitutional monarchies. The region's monarchs have varying levels of power. The sultan of Brunei has almost complete power, while Cambodia's king is limited by a democratically elected legislature.

Culture and Society

Maritime Southeast Asia's location at the crossroads of important trade routes led to its being influenced by a variety of cultures. The region has added new ideas to its indigenous cultural traditions.

Education and Healthcare

On the islands of Southeast Asia, education levels have increased since independence. Most children in Indonesia now attend primary school. In Indonesia and Malaysia, the literacy rate is about 88%. In the Philippines, Singapore, and Brunei, the literacy rate is about 93%.

Healthcare is better on the islands than the mainland of Southeast Asia. In Indonesia and Malaysia, the government provides most modern healthcare services, although they are usually better in cities. Singapore states that the quality of its healthcare is as good as that of developed countries.

Language and Religion

Indonesia has over 300 ethnic groups with over 250 distinct languages. Colonization added new languages to the islands of Southeast Asia. English and Spanish are common in the Philippines. Due to the importance of global trade, Singapore's official languages are Chinese, Malay, Tamil, and English.

The practice of Islam is widespread across the islands of Southeast Asia.

> **DID YOU KNOW?**
>
> Indonesia has the largest population—202,867,000—of Muslims in the world according to a 2009 study by the Pew Research Center.

As a result of Spanish colonization, most people in the Philippines are Roman Catholic. Many Southeast Asians of Chinese descent follow Confucianism or Taoism.

Arts

The ancient art and architecture of India and China influenced Southeast Asia. The stunning Buddhist shrine of Borobudur in Indonesia was built of gray volcanic stone around A.D. 800. Each level of the shrine is connected by stairs that symbolize the stages of the Buddha's journey to enlightenment. The soaring Petronas Towers of Kuala Lumpur, Malaysia, blend traditional and modern styles of architecture.

Traditional dances of Southeast Asia often use religious themes. On the Indonesian island of Bali, young women perform a dance called the *Legong*. Puppet plays using historical and religious characters to tell tales are popular in many parts of Southeast Asia.

Leisure and Celebrations

Popular culture in Singapore is based on modern mass media such as films from Hong Kong, Taiwan, and the United States. Badminton and soccer are popular in Indonesia.

Religious festivals and state and national holidays are celebrated throughout Southeast Asia. In Indonesia, most government and private organizations observe national holidays such as the Chinese New Year and the **Prophet** Muhammad's birthday.

Practice

Fill in the blank with the correct word from the information in the preceding paragraphs.

1. ___________ has the largest Muslim population in the world.

2. ___________ is the most modern and crowded country in Southeast Asia.

3. Following European colonization, ___________ occupied the islands of Southeast Asia.

4. Europeans changed the economies of the islands from small farms to large ___________ plantations

5. India brought the Hindu and ___________ religions to Southeast Asia.

6. ___________ cities like Bangkok serve a country's ports and economic centers, and are often the capital city.

7. Arab traders brought the religion of ___________ to Southeast Asia.

8. The ___________ Rouge were harsh communist rulers in Cambodia.

9. Southeast Asian civilizations developed on ___________ or around strategic ports.

10. ___________ has always been the main staple crop of Southeast Asia

Economy

Southeast Asia's fertile soil, warm climate, and rich mineral deposits affect the region's economic activities. War and political changes have affected economic growth in Southeast Asia. Still, some countries are rapidly developing and industrializing.

In the past 20 years, Vietnam's government loosened economic control. Its economy has made notable progress through *moi*, or renovation.

Many countries in Southeast Asia are industrializing. Still, most people continue to make their living in agriculture. More than 75% of people in Laos are farmers.

Rice Cultivation

More than half the region's arable land is used to grow rice. Vietnam, Indonesia, Thailand, and Myanmar are its top rice producers. Thailand and Vietnam are the world's leading rice exporters. Rice grows well in Southeast Asia because of the abundant water supply, fertile volcanic and **alluvial soil**, and warm, wet climate.

Seasonal flooding of the Mekong and Chao Praya Rivers irrigate **rice paddies**, the flooded fields where rice grows. Rainwater allows rice to grow in Myanmar's Irrawaddy River delta and in parts of the Philippines.

Rice farming is difficult without modern machinery. Most farmers plant and harvest crops by hand using simple tools such as long, sharp, curved knives called **sickles.** Water buffalo or oxen are traditionally used to pull plows, but some farmers now use engine-powered plows.

Other Crops

In drier areas of Southeast Asia, yams, corn, bananas, and cassava are grown. Indonesian farmers grow the edible cassava root because it is easier to grow than rice.

Plantations on the coastal lowlands provide cash crops.

DID YOU KNOW?

Indonesia, Thailand, and Malaysia lead the world in rubber production.

On a stretch of land 700 miles long on each side of the equator called the rubber belt, large plantations grow the world's best rubber trees. Java and the Philippines grow sugarcane. Regional exports include coffee, palm oil, coconuts, and spices.

Fish Farming

Since people in Southeast Asia consume twice the world average in seafood, fish remains an important part of the regional economy. Large fleets of trawlers compete with traditional fisherman and small operations. Fish yield has increased, but overfishing became a concern. World demand for seafood has decreased, so excessive fishing has leveled off.

Forestry and Mining

Forestry (logging, transporting timber, and manufacturing finished wood products) is important to many Southeast Asian countries. Teak, ebony, and bamboo lumber and finished products are essential to the economies of Malaysia, the Philippines, Indonesia, and Thailand. Logging has increased deforestation. Many countries are working to make economic and environmental goals compatible.

There are rich mineral deposits beneath the mountains of Southeast Asia. Malaysia, Thailand, and Indonesia are leading producers of tin. Malaysia and the Philippines extract iron ore.

Malaysia and Indonesia have large petroleum and natural gas reserves. Almost half of Brunei's export income comes from crude oil, natural gas, and petroleum products.

DID YOU KNOW?

Indonesia is Southeast Asia's largest petroleum producer. It was the only Asian member country in OPEC until it withdrew in 2008.

The Indonesian government has reserved large areas of Papua for resource development. Located on the western side of New Guinea, Papua has timber resources and rich mineral **lodes**, or deposits. Most of the people in Papua are poor. Proponents of independence claim the Indonesian government has allowed foreigners to extract resources while investing very little to improve health, education, and public services.

Transportation and Communications

Methods of transportation and distribution of communications vary across Southeast Asia. Transportation and communications are affected by physical features, industrialization, and economic development.

Southeast Asia's rivers, long coastlines, islands, and peninsulas make water transportation the most common form of transport for people and goods. Overflowing rivers can make travel difficult.

Southeast Asia has long been the crossroads of major ocean routes. Most shipping today between Europe and East Asia passes through the Strait of Malacca near Singapore.

DID YOU KNOW?

Singapore's strategic location makes it a prosperous **free port** where goods can be unloaded, stored, and reshipped without paying import taxes, tariffs, or duties.

Haiphong, Vietnam; Bangkok, Thailand; Jakarta, Indonesia; and Manila, Philippines are also important regional ports.

Because of different levels of economic development, the quality of land transportation varies across Southeast Asia. Cambodia lacks resources to build an effective highway network. With more successful economies, Singapore, Malaysia, and Thailand can fund road improvements and build urban rail systems.

Highways and railroads on Southeast Asia's peninsulas and large islands only link major cities. Paved roads in urban centers are overflowing with trucks, automobiles, motorcycles, and buses. In most of rural Southeast Asia, rugged terrain, forests, and seas separating islands make travel difficult.

People and Environment

Industrialization and economic development in Southeast Asia have polluted the air, land, and water, and destroyed valuable natural resources. Population growth and urbanization have contributed to environmental problems in the region. Poverty and political conflict make these challenges more difficult as people are forced to choose between personal survival and environmental conservation.

Urban Environments

Southeast Asia's increased prosperity raised people's quality of life expectations, but economic growth siphons off limited resources. Industries and manufacturing raise standards of living and create industrial waste.

TIP

Growing populations and crowding in Bangkok, Jakarta, and Manila raise housing, sanitation, water, and traffic control issues.

Bangkok is a bustling factory city, with skyscrapers and expressway traffic jams. The large population increases and expansion of industry have raised Bangkok's temperature, humidity, and pollution levels faster than the global average. Health problems are on the rise in Bangkok.

Rural Environments

Pollution extends into rural areas and national parks in parts of Southeast Asia. Poor waste disposal contaminated most of the freshwater wells in one of Thailand's national parks. The dumping of toxic waste is a problem.

The clearing of land for farming and the timber industry sometimes causes forest fires. Indonesian forest fires have sent a haze of dust, ash, sulfur dioxide, and carbon dioxide into the air and caused respiratory problems for people as far away as mainland Malaysia.

Managing Resources

Southeast Asia's natural resources are economically important, but governments must find ways to develop their economies without further destroying resources.

Minerals, metals, and rain forest timber are Southeast Asia's most valuable resources and sources of income.

Reforestation and Conservation

Southeast Asian countries have recently taken steps to protect their environments. To stop further loss of rain forests, Malaysia, the Philippines, Indonesia, and Thailand have limited certain timber exports and started **reforestation** programs. Enforcement has been a problem and illegal logging is reducing forests. A variety of biodiverse environments and plant and animal species are in danger of being lost within a few years.

In the 1980s, Indonesia planned to set aside large portions of the country as environmental conservation areas. The government abandoned the plan and granted logging rights to timber companies amid political turmoil.

Southeast Asian governments are beginning to deal with the impact of urban growth on the environment.

DID YOU KNOW?

Because of industrialization, overcrowding, and expanded use of automobiles and other vehicles, Bangkok, Thailand, is called an urban heat island.

Possible solutions to this problem include creating "green zones" of environmental protection within cities. Another idea is banning the construction of tall buildings near the sea so winds can blow into the city and provide ventilation.

Human Impact

Human activities have had a negative impact on the environment, natural resources, and wildlife of Southeast Asia. Industrialization, population growth, and economic development have caused pollution and the destruction of natural resources. Whether commuting through traffic in a busy city or controlling floodwaters to protect a cash crop, Southeast Asians are challenged to work in harmony with their environment.

Timber and Agriculture Industries

Malaysia, Thailand, Laos, and Myanmar rely on teak and other timber for income. Since the 1960s, commercial logging companies have set up modern logging processes and provided training and jobs for many Southeast Asians. Economies have benefited, but the region's forests have been diminished.

Mining Industry

The mining of metals and minerals has led to environmental abuses. At Indonesia's largest gold mine waste is dumped into the Akjwa River in Papua. Dumping diverts the river from its course and displaces people.

Fishing Trade

Coral reefs, like tropical rain forests, have amazing biodiversity. Local people use poisons and explosives to capture certain types of fish in demand at Asian restaurants and the world's aquariums. Live reef-fish trade generates huge profits, but destroys the reefs. Once the reefs are gone, local communities lose their primary food source.

Shipping and Trade

For centuries, the lower Mekong River has been one of Southeast Asia's major water highways. Shallow water and rapids have made travel in the upper Mekong dangerous. In 1992, China started the Upper Mekong Navigation Improvement Project, which is deepening and widening the river and building dams upstream to increase the river's shipping capacity. The navigation plan should bring needed economic benefits, but critics wonder about the impact on local fisheries and agriculture.

Future Challenges

Due to Southeast Asia's location, countries in the region face continuing challenges from natural disasters.

Floods and Typhoons

Each year in Southeast Asia, hundreds of people are killed and millions of crop acres are destroyed by storm-induced flash floods. Human activity magnifies the effects of floods—in the Philippines in 1991 and 1995, so much forest had been cleared there were widespread runoff and mudslides.

The rivers of mainland Southeast Asia have seasonal flooding each year. Built on unstable land, Bangkok is facing a constant threat of flooding.

DID YOU KNOW?

Some parts of Bangkok are sinking more than 25 inches a year.

Mount Pinatubo and Evacuation

The eruption of Mount Pinatubo in the Philippines in June 1991 killed more than 300 people and caused damage up to 11 miles away. Volcanic fallout rendered 212,511 acres of farmland and fishponds infertile. More than 100,000 people were left homeless. Healthcare facilities, airports, schools, and a U.S. air base were damaged and closed. Fine particles called aerosols were released into the upper atmosphere; the haze lasted for three years. Weather patterns across the globe were affected. Global temperatures dropped nearly 1°F and the amount of sunlight reaching Earth was reduced.

Scientists had predicted Mount Pinatubo's eruption and the authorities were prepared. Effective monitoring of the volcano prior to eruption allowed for the government's successful evacuation of 60,000 Filipinos and 18,000 American military workers and their families at a nearby base. The evacuation saved thousands of lives as mud slides, a common occurrence following volcanic eruptions, engulfed nearby villages.

Other Volcanic Eruptions

Bali's Mount Agung erupted in 1963. The Balinese people worship the volcano as the sacred hallmark of their faith and leave food and flower offerings on the crater's rim. Despite the eruption that took over 1,500 lives, many Balinese people risk their lives and property to live near Mount Agung.

Practice

Insert T or F to indicate whether the statement is true or false, based on what you've read.

11. ____ The Vietnamese government still strictly controls the country's economy.

12. ____ Southeast Asia is a leading producer of rubber.

13. ____ Indonesia lacks valuable fossil fuels.

14. ____ Most shipping today between Europe and East Asia passes through the Strait of Hormuz near Singapore.

15. ____ Healthcare is better on mainland Southeast Asia than on the islands.

16. ____ Minerals, metals, and rain forest timber are Southeast Asia's most valuable resources and sources of income.

17. ____ Sustainable development is not important for natural resource management.

18. ____ Creating "green zones" and limiting tall building near coastlines may help urban heat islands like Thailand.

19. ____ Economic growth siphons off limited resources.

20. ____ Air and water pollution, deforestation, and toxic waste dumping are naturally occurring problems in Southeast Asia.

Answers

1. Indonesia
2. Singapore
3. Japan
4. Commercial
5. Buddhism
6. Primate
7. Islam
8. Khmer
9. Waterways
10. Rice
11. F
12. T
13. F
14. F
15. F
16. T
17. F
18. T
19. T
20. F

LESSON

23 AUSTRALIA AND NEW ZEALAND, OCEANIA, AND ANTARCTICA—PHYSICAL GEOGRAPHY

LESSON SUMMARY

Australia, Oceania in the South Pacific, and Antarctica are extremely different. The Australian Outback is dry. In the South Pacific, there are volcanic islands. Antarctica is a cold ice cap. Each of these subregions offers unique opportunities for economic growth, tourism, and scientific research.

For we are as connected to the planet as the corals are to the algae living within them.

—Jean-Michel Cousteau, in the 2003 documentary *Coral Reef Adventure*

Australia

Australians have adapted to life in a country with large expanses of dry, flat land. At the same time, it is surrounded by water like an island. Australia's physical environment contributes in various ways to the country's economy.

DID YOU KNOW?

Australia is the only place in the world that is a continent and country. It is the smallest of the continents, but the sixth-largest country.

Mountains and Plateaus

The Great Dividing Range, also known as the Eastern Highlands, is a chain of hills and mountains interrupting Australia's flat landscape. Peaks stretch more than 2,300 miles along Australia's east coast from the Cape York Peninsula to the island of Tasmania. Elevations range from 2,000 feet to more than 7,000 feet. The southern highlands, averaging 3,000 feet, are known as the Australian Alps. Mount Kosciuszko, at 7,310 feet, is Australia's tallest mountain. Most of Australia's rivers start in the range and water the country's most fertile land.

The Western Plateau is a primarily low flatland in central and western Australia covering two-thirds of the continent. Few people live in this Outback. Erosion caused the Western Plateau's barren landscape. Across the plateau are the Great Sandy, Great Victoria, and Gibson deserts. A few erosion-resistant mountains exist in the Western Plateau, including the Hamersley, MacDonnell, and Musgrave mountain ranges. Many gorges are also in this region.

South of the Great Victoria Desert is the virtually treeless Nullarbor Plain. The plain ends abruptly in the immense Bunda Cliffs. The Koonalda Caves are among the many limestone caves in this area. Churning hundreds of feet below the cliffs is the Great Australian Bight, a bay.

DID YOU KNOW?

The Nullarbor Plain is the world's largest single piece of limestone. It measures roughly 77,000 square miles.

Central Lowlands

The arid grassland and desert of the Central Lowlands in east central Australia separate the Great Dividing Range from the Western Plateau. Lakes and rivers in the Central Lowlands are dry most of the year. When it rains heavily, the rivers and lakes fill with water.

In the southeast, the Murray and Darling rivers provide water for irrigation. The Murray River is Australia's longest river at over 1,400 miles. It rises in the Australian Alps and empties into Lake Alexandrina. The Darling is 915 miles long and flows through Southern Queensland. The Great Artesian Basin underlies one-fifth of the continent. Waters from the Eastern Highlands are absorbed by the lowlands in underground wells, some over a mile deep. More than 350 million gallons of water are discharged to the surface daily. Ranchers use these **artesian wells** for livestock since it is too salty for humans or crops.

Great Barrier Reef

The world's largest coral reef, the Great Barrier Reef, lies along Australia's northeast coast in the Coral Sea. It reaches from the Torres Strait in the north to Lady Elliot Island in the south. Warming temperatures and rising sea levels created conditions for the formation of the reef. The reef is a national park and a United Nations World Heritage site because of its beauty and the habitats it provides to thousands of plant and animal species. The reefs extend more than 1,500 miles. Several islands surround the reef with an abundance of plant and bird life and new luxury resorts. Coral bleaching, a loss of pigment in the coral that signifies stress or death, is increasing due to rising temperature of the water.

DID YOU KNOW?

The Great Barrier Reef is a chain of more than 2,500 small reefs formed from the limestone skeletons of a tiny sea animal.

Natural Resources

Agriculture is important to Australia even though only 10% of its land can be farmed. Farmers grow wheat, barley, fruit, and sugarcane along the Murray and Darling rivers. Ranchers raise cattle, sheep, and chickens in the arid Outback.

Australia is mineral rich. With deposits of petroleum, coal, iron ore, lead, zinc, gold, and nickel, Australia is a world leader in mining. Australia has most of the world's high-quality opals.

DID YOU KNOW?

Australia has over 25% of the world's raw aluminum, bauxite.

Oceania

Oceania is made up of thousands of islands of different sizes spread across millions of miles of the Pacific Ocean. Some islands formed when tectonic plates collided millions of years ago. Other islands were created by volcanic hot spots. Life on the islands of Oceania is influenced by the type of island and the physical processes that formed it.

Island Groups

The islands of Oceania can be classified by location, how they formed, and the culture of inhabitants. The "black islands" of **Melanesia**, including New Guinea and Fiji, lie north and east of Australia. The 607 "little islands" of **Micronesia** lie north of Melanesia. The "many islands" of **Polynesia** cover an area larger than Melanesia and Micronesia. They include Samoa, the Cook Islands, French Polynesia, and others. Polynesia extends from Hawaii's Midway Island in the north to New Zealand in the south. The Polynesian Triangle refers to the shape formed by the island groups of New Zealand, Easter Island, and Hawaii, and all the islands that fall within it, such as Samoa, the Cook Islands, and French Polynesia.

Island Types

Volcanic eruptions and earthquakes occur on high island landscapes. High islands such as Tahiti have mountain ranges split by valleys that fan out into coastal plains. The volcanic soil on high islands supports some agriculture. Freshwater bodies dot the land.

Volcanoes also shaped Oceania's low islands. Many of the Marshall Islands are ring-shaped low islands called **atolls**. Coral reef buildup on the rims of submerged volcanoes makes atolls. Atolls encircle shallow pools of clear water called **lagoons** that rise just a few feet above sea level.

The rising and folding of ancient rock from the ocean floor creates continental islands from bodies of land that lie on the continental shelf.

TIP

Most of Oceania's larger islands, like New Guinea and New Caledonia, are continental islands.

Coastal areas include swamps, rivers, and plains. Interiors include volcanoes, mountains, plateaus, and valleys. Continental islands have a variety of rocks and soil and contain most of Oceania's minerals. Continental mining yields oil, gold, nickel, and copper. Some larger forested islands process timber.

New Zealand

The sandy beaches, emerald hillsides, and snow-tipped mountains of New Zealand's North Island and South Island lie 1,200 miles southeast of Australia, across the Tasman Sea. Cook Strait separates the two mainland islands. The northern part of North Island has golden beaches, ancient forests, and rich soil. The central plateau of volcanic stone has hot springs and several active volcanoes including North Island's highest point, Mount Ruapehu. The plateau has many shining freshwater lakes. East of the plateau, a band of hills runs north and south.

On South Island's western edge stand the towering snow-capped peaks of the Southern Alps and the capital city of Wellington.

DID YOU KNOW?

New Zealand's earliest inhabitants, the Maori, named the highest peak on South Island Aorangi, meaning "cloud piercer."

Also called Mount Cook, Aorangi rises 12,316 feet amid glacier-carved, sparkling lakes and tumbling rivers. New Zealand's flattest and most fertile land, the Canterbury Plains lie on the east coast. The west coast has rugged cliffs, deep fjords, and coastal caves. The north is thickly forested.

New Zealand's most valuable natural resource is its fertile volcanic soil. Half of the land supports crops and livestock. Sheep and wool products are New Zealand's top exports. Its forests yield valuable timber. New Zealand's rivers provide abundant hydroelectric power.

DID YOU KNOW?

New Zealand uses geothermal energy—water heated underground by volcanoes to generate power.

Most of the hundreds of New Zealand's outlying small islands are populated. Stewart, Chatham, Great Barrier, and Auckland are some of the larger small islands of New Zealand. The land mass of all of the small islands and North Island and South Island together is approximately equal to the size of the state of Colorado.

Antarctica

Antarctica is a continent located at the southernmost point on Earth. It is surrounded by the Pacific, Atlantic, and Indian Oceans. Its nearly 5.4 million square miles is completely covered by ice. Antarctica has no indigenous people and very limited plant and animal life. Since Antarctica was first seen by modern human eyes in 1820, the frigidly cold continent has intrigued explorers and scientists.

DID YOU KNOW?

Mount Erebus, on Antarctica's Ross Island, is the world's southernmost active volcano.

Land and Climate

The Transantarctic Mountains separate this vast ice-covered land into East Antarctica, a landmass about the size of Australia, and West Antarctica, a group of islands. East Antarctica contains the South Pole and is covered by a massive ice dome rising from coastal plains to a high plateau. West Antarctica is geographically younger and has several active volcanoes. Gravity-driven winds, known as katabatic winds, blow from the high interior to the coast. **Blizzards** often form in the plateau.

DID YOU KNOW?

Vinson Massif in West Antarctica, at 16,066 feet, is the highest point in all Antarctica.

Antarctica's climate is cold and the severity depends on location. East Antarctica is higher in elevation and the coldest.

DID YOU KNOW?

In East Antarctica, the lowest yearly temperature is about –126°F.

The Antarctic Peninsula and surrounding islands of West Antarctica have the mildest temperatures. During the heart of Antarctic summer in January, the average temperature is above freezing.

Plant and Animal Life

Plant and animal life are difficult to sustain in the freezing temperatures and isolation of Antarctica. Most plant life is found in the milder climate of West Antarctica, especially the Antarctic Peninsula. There are no trees in Antarctica; algae, lichens, and mosses make up the vegetation.

Most animal life in Antarctica can be found in the Antarctic Ocean. Penguins, whales, and seals survive on a variety of food sources including krill, larger fish, and squid.

The Antarctic minke whale has been hunted for its blubber, meat, and oil. International treaties prohibiting hunting protect the remaining minke whale population, under threat from hunters since the early nineteenth century.

DID YOU KNOW?

Antarctica is home to some 70 lakes, including Lake Vostok, which may contain microbial life.

The emperor penguin is well adapted to the Antarctic environment. The emperors swim and catch food in the water, but lay their eggs and hatch their young on land.

Antarctic hair grass and pearlwort are the only two flowering plants found on Antarctica. They bloom and set seeds very quickly during the Antarctic spring.

Natural Resources

Scientific evidence shows the continent is rich in coal, copper, lead, zinc, silver, gold, oil, and natural gas. The harsh conditions in Antarctica make resource extraction difficult. The environmental and political impacts of mining in Antarctica are a deep concern. Mining restrictions in Antarctica were laid out in the 1991 Protocol on Environmental Protection.

DID YOU KNOW?

Only 1% of Antarctica's land has been surveyed for minerals.

Exploration and Scientific Research

In the 1770s, Captain James Cook became the first modern person to cross the Antarctic Circle. Captain Cook did not see land, but throughout the nineteenth century explorers, scientists, and whalers discovered land. In 1821, American seal hunters made the first known landing on Antarctica.

In 1901, British explorer Robert F. Scott began the first known inland exploration of Antarctica. In 1911, Norwegian Roald Amundsen became the first human known to reach the South Pole. In 1929, Admiral Richard E. Byrd made the first flight over the South Pole.

The International Geophysical Year of 1957–1958, a year of global geophysical research, focused world attention on Antarctica. New bases were established and new research was carried out.

DID YOU KNOW?

In 1959, the 12 countries involved in the project negotiated the Antarctic Treaty to preserve Antarctica for peaceful scientific research. All territorial claims were put on hold.

Since 1959, other countries have established research programs in Antarctica and 26 new countries have signed on as consultative members. There are 43 stations operated by 27 countries. Station populations are about 4,000 in summer and about 1,000 in winter.

Tourism

In 1958, the first tourists visited Antarctica. Tourism in Antarctica is slowly growing; in the 1990s, about 10,000 tourists visited. Most of these tourists took cruises from Australia or New Zealand and only made short excursions onto the mainland. In 2005 and 2006, the number of tourists grew to almost 30,000. Some people are concerned about human effects on the environment of Antarctica, but most people think tourism will increase interest in the region. The cold temperatures make the tourist season short. It takes place during the Antarctic summer, which runs from October to February.

Practice

Fill in the blank with the correct word from the information in the preceding paragraphs.

1. The Canterbury Plains are ___________'s flattest and most fertile land.
2. Australia's cattle ranching Western Plateau is also called the ___________.
3. The world's largest coral reef, located off the coast of Australia, is the ___________ ___________ Reef.
4. British explorer Robert F. Scott began the first known inland exploration of ___________ in 1901.
5. Only 1% of the land in Antarctica has been surveyed for ___________.
6. The ___________ and ___________ are two animals that can be found in the Antarctic Ocean.
7. The country of ___________ harnesses underground geothermal energy created by volcanic heat.
8. Polynesia, Melanesia, and Micronesia make up ___________.
9. ___________ encircle shallow pools of clear water called lagoons that rise just a few feet above sea level.
10. Tahiti is a ___________ island.

Climate and Vegetation

Australia has several different climates. Oceania has a tropical wet climate. Australia and Oceania experience wet and dry monsoons each year. Such seasonal weather patterns impact the natural vegetation of the region and the lives of people. New Zealand has a marine west coast climate. Each type of climate affects human activities.

Australia

In Australia, differences in rainfall cause differences in climate and vegetation. In Australia, there are tropical climates in the north, deserts in the interior, and midlatitude temperate areas of grassland, scrub, and mixed forests along the eastern, southern, and southwestern coasts.

Each year from December to March, subtropical, high pressure air masses block moisture-laden Pacific Ocean winds from reaching the Western Plateau, Australia's large interior desert area. The sun scorches the land, but nighttime temperatures drop dramatically.

A milder steppe climate encircles Australia's desert region. In the steppe, more regular rainfall results in vegetation including eucalyptus, acacia trees, and small shrubs.

DID YOU KNOW?

Acacia tree saplings were used by early settlers to make **wattle**, a strong interwoven wooden framework used for home construction.

The coastal areas of Australia have a variety of moister climates. The humid subtropical northeastern coast averages more than 80 inches of rain a year. Less rain falls in the Mediterranean climate area of the southern coasts and in the marine west coast climate area along the southeastern coast. These coastal areas provide Australia with most of its agriculture.

Oceania and New Zealand

Since most of Oceania lies between the equator and Tropic of Capricorn, many of the islands have a tropical wet climate. Most days are warm and range from 70 to 80°F.

Seasons in most of Oceania alternate between wet and dry. The dry season has blue skies without clouds and the wet season has constant rain and high humidity. Low islands receive little rainfall. The larger landmasses of high islands give off warm, moisture-laden air. When this air rises to mix with cool ocean breezes, there is heavy rainfall.

Dry, low islands only have shrubs and grasses. Islands with more rainfall have coconut palms and other trees. Hot, steamy rain forests thrive where heavy rains drench island interiors. A mostly windless area called the doldrums is a narrow band near the equator where opposing ocean currents meet. The calm within the doldrums can turn into violent storms known as typhoons.

Most of New Zealand has a more temperate marine west coast climate. In winter, ocean winds warm the land. The winds from the ocean cool New Zealand in summer. Temperature extremes are controlled by the ocean. In summer, temperatures range from 65 to 85°F. In winter, temperatures range from 35 to 55°F.

Climate variations result from geographic differences. The central plateau of North Island is warm and sunny during summer, but the mountaintops may have snow all year. Mountainous areas exposed to western winds often have more rainfall than other areas. Although New Zealand averages 25 to 60 inches of rain a year, the Southern Alps on South Island have an average annual rainfall of 315 inches.

Because of its geographic isolation, New Zealand has unique plant life. About 90% of New Zealand's indigenous plants are native only to New Zealand. A small shrub called **manuka** grows where prehistoric volcanic eruptions destroyed ancient forests. Early settlers from Great Britain cut down almost all the pinelike kauri trees. Some still grow amid thriving evergreen forests. Several tree species have been imported to address erosion in deforested areas.

Practice

Insert T or F to indicate whether the statement is true or false, based on what you've read.

11. _____ The doldrums are calm and do not produce typhoons.

12. _____ New Zealand has two major islands and a marine west coast climate with variations.

13. _____ The northern regions of Australia enjoy a tropical climate.

14. _____ Winds from the ocean cool New Zealand in summer and warm it in the winter.

15. _____ Islands with more rainfall have coconut palms and other trees.

16. _____ Australia has desert, steppe, and forest vegetation.

17. _____ Seasons in most of Oceania alternate between wet and dry.

18. _____ Most of New Zealand's indigenous plants can only be found there.

19. _____ In the Australian steppe, more regular rainfall results in vegetation like eucalyptus, acacia trees, and small shrubs.

20. _____ Oceania has a tropical wet climate.

Answers

1. New Zealand
2. Outback
3. Great Barrier
4. Antarctica
5. Minerals
6. Minke whale, penguin, seal, squid, and fish.
7. New Zealand
8. Oceania
9. Atolls
10. High
11. F
12. T
13. T
14. T
15. T
16. T
17. T
18. T
19. T
20. T

LESSON

24 AUSTRALIA AND NEW ZEALAND, OCEANIA, AND ANTARCTICA—HUMAN GEOGRAPHY

LESSON SUMMARY

The geography and climates of Australia, Oceania, and New Zealand have drawn people from great distances. Despite physical barriers and long distances for those who live in and visit Australia and Oceania, improved transportation and communications have made the region more interdependent.

We are all visitors to this time, this place. We are just passing through. Our purpose here is to observe, to learn, to grow, to love . . . and then we return home.

—Australian Aboriginal Proverb

Countries

Australia
Fiji
Kiribati
Marshall Islands
Micronesia, Federated States of
Nauru
New Zealand
Palau
Papua New Guinea
Samoa
Solomon Islands
Tonga
Tuvalu
Vanuatu

Australia and New Zealand

Migration and settlement patterns have influenced the cultures and landscape of Australia and New Zealand. For thousands of years, Australia and New Zealand have been meeting places. Dramatic deserts, mountains, and forests meet oceans on temperate coastlines. In the colonial period, British colonists met indigenous people. Travelers today can meet amazing wildlife in Australia and New Zealand.

Population Patterns

Indigenous peoples and foreign colonizers influenced the look of modern-day Australia and New Zealand.

People

Australia's earliest settlers, the **Aboriginal** people may have the world's oldest surviving culture. More recent arrivals provide Australia with great diversity. These nomadic hunters and gatherers arrived in Australia 40,000 to 60,000 years ago from Southeast Asia. About 2% of the population of Australia is Aboriginal.

DID YOU KNOW?

Aboriginal people feel a direct relationship with the landscape and believe in Dreamtime, a system of beliefs that connects them to the beginning of time.

Different groups live in different regions of the continent. The Arrente have lived in central Australia for 20,000 years. The Palawa have lived on the island of Tasmania for about 32,000 years.

The **Maori** of New Zealand came from the islands of Polynesia. The Maori have lived by hunting, fishing, and raising crops. Many ancient Maori traditions still survive.

Beginning in the 1500s, Europeans started sailing the waters around Australia and New Zealand. Europeans eventually colonized the region. Today, most of the people in Australia and New Zealand are of British descent.

Australia has recently recruited immigrants. The number of East and Southeast Asians who have moved to Australia for economic opportunities has increased.

Density and Distribution

Australia's physical geography results in uneven distribution of people. Very few people live in the dry central plateaus and deserts. Most people live along the southeastern, eastern, and southwestern coasts where there are mild climates, fertile soil, and access to the sea. Most of New Zealand's people live along the coasts as well.

The largest Australian cities are Sydney and Melbourne. Each has more than 3 million residents and they serve as major commercial ports. New Zealand's ports of Auckland, Christchurch, and Wellington are the country's largest cities.

History and Government

Early indigenous inhabitants, the effects of British colonization, and the experience of independence have shaped the cultures of Australia and New Zealand.

Early Peoples

Australia's earliest settlers may have migrated over land bridges during the Ice Age when ocean levels were much lower than today. The early Aboriginals were nomads using well-traveled routes to reach water and seasonal food sources. Family groups called clans traveled together in their ancestral territories, carrying only baskets, bowls, spears, and sticks for digging.

> **DID YOU KNOW?**
>
> The term *Aboriginal* refers collectively to the many indigenous communities and societies of mainland Australia.

Increased trade expanded migration across the islands. Between A.D. 900 and 1300, the Maoris left eastern Polynesia and settled in New Zealand. Maori farmers lived in villages and grew traditional root crops like taro and yams that they brought from their Polynesian homeland.

European Exploration

From the 1500s to the 1700s, Europeans from various nations explored the vast spaces of the South Pacific. British Captain James Cook led three voyages to the region between 1768 and 1779. In 1769, he became the first European to set foot on New Zealand. In 1770, he discovered what would become known as the Great Barrier Reef. When he arrived that same year on the eastern coast of Australia, he claimed the land for Great Britain.

European Settlement

Beginning in 1788, Great Britain started using Australia as a penal colony to which convicts from overcrowded British prisons were transported. By the 1850s, the imprisonment of British convicts ended and free British settlers were establishing coastal farms and settlements. Livestock, mostly sheep, were introduced to the continent. Settlers profited from the export of wool to Britain. Gold was discovered in Australia in the 1850s and it has been a source of wealth ever since.

Attracted by the rich soil and fishing grounds, the British and other Europeans started establishing settlements in New Zealand. By the end of the 1800s, raising livestock was a major segment of New Zealand's economy.

Indigenous Peoples

The arrival of Europeans had a disastorous impact on indigenous peoples. The British forcibly removed many Aborigines from the land and denied them basic rights. Diseases and violence steadily reduced their populations. Still, the Aboriginal people fought back. In the mid-1800s, Aboriginals were placed in segregated reserves.

British settlements in New Zealand brought hardships for the Maori. When the British colonists introduced new ways of farming and European culture, the Maori social structure weakened. In the 1800s, the Maoris armed themselves and resisted British rule for 15 years. Many Maori were killed and most of their land was lost to the British.

> **DID YOU KNOW?**
>
> Before European settlement in Australia, there were more than 200 Aboriginal languages and about 600 dialects. Today, most Aboriginal people speak English.

Independence

In the early 1900s, Australia and New Zealand peacefully won their independence from Great Britian. In 1901, Britain's Australian colonies became states and formed the Commonwealth of Australia. The new country was a dominion, a largely self-governing country within the British Empire. In 1907, New Zealand became a self-governing dominion with a British parliamentary system.

DID YOU KNOW?

In 1893, New Zealand became the first country in the world to legally recognize women's right to vote.

Since World War II, Australia and New Zealand have forged closer economic and political relationships with the United States. The Aboriginals and Maori have won greater recognition of their unique cultural identities and become politically active.

Culture and Society

Indigenous cultures and European traditions shaped the cultures of Australia and New Zealand. Asian influences have also added to the ethnic mix. Although daily life in New Zealand and Australia looks a lot like it does in other places in the West, there are many different ways of life in different parts of the region.

Education and Healthcare

The quality of education varies throughout Australia and New Zealand. Both countries provide free, compulsory education, and literacy rates are over 99%. Many students attend universities. Many students in Australia's remote Outback receive and turn in assignments by mail or communicate with teachers via two-way radios. In recent years, the Internet has led to increased and faster communication, with courses and information available all online.

Australians and New Zealanders, particularly in cities, usually have access to quality medical care and other social services. In some parts of Australia, rugged terrain and long distances make access to healthcare difficult. However, doctors consult with patients through the use of two-way radios and mobile clinics.

Indigenous people do not generally receive such benefits. Aboriginal people suffer from poverty, malnutrition, and unemployment. Recently, the Australian government and private organizations have been trying to make up for past injustices. Courts have recognized the claims of Aboriginal people to government assistance, land, and natural resources.

Language and Religion

English is the major language in Australia and New Zealand.

DID YOU KNOW?

Australian English, called **Strine**, has a unique vocabulary made up of Aboriginal words and slang created by modern Australians.

Because the Maori population of New Zealand is quite large, their language is widely spoken in certain areas. Only about 2% of Australians, about the number of Aboriginal people in the country, speak Aboriginal languages.

The religion of the indigenous people of Australia and New Zealand focuses on the connection between humans and nature.

Europeans brought Christianity, which attracted many indigenous followers. Today, Christianity is the most widely practiced religion in Australia and New Zealand.

Arts and Leisure

The peoples of New Zealand and Australia traditionally used art, music, dance, and storytelling to pass on knowledge from generation to generation. Australian Aboriginals recorded their past in rock paintings and developed songs to pass on information.

Sports and leisure in Australia and New Zealand reflect the region's colonial heritage. British settlers brought cricket to Australia and New Zealand. In urban areas, where Western influence is dominant, leisure activities, including tennis, boating, waterskiing, and other water sports appear alongside metropolitan beaches.

Oceania

Indigenous and Western cultures have shaped the societies of Oceania. Hundreds of cultures on thousands of islands in the South Pacific have shared religious beliefs that tie them to the land and sea.

Population Patterns

Migrations of people among the islands in Oceania have shaped life on the islands today.

Many Peoples

People came to Oceania from Asia more than 45,000 years ago. Waves of immigrants from Asia continued to arrive over the centuries. People already living there moved from island to island and settled into three major groups—the Polynesians, Micronesians, and Melanesians.

Melanesia

Located in the southwest Pacific Ocean, Melanesia includes the independent island countries of Papua New Guinea, Fiji, and the Solomon Islands, as well as French-ruled New Caledonia. Melanesian groups, even those on the same island, differ greatly. The Chimbu, the largest indigenous group of Papua New Guinea, is known for its egalitarian social structure.

Micronesia

Micronesia is in the western Pacific east of the Philippines. The independent countries of Micronesia include the Federated States of Micronesia, Nauru, and Kiribati.

> **DID YOU KNOW?**
>
> Micronesia includes the U.S. territories of Guam and the Mariana Islands. Polynesia includes the U.S. state of Hawaii.

Micronesians have several ethnicities, all derived from the Micronesian culture. More than a dozen indigenous languages are spoken by Micronesians. Micronesia has four official national languages, of which English is one.

Polynesia

Polynesia is located in the central Pacific Ocean. The three independent countries of Polynesia are Samoa, Tonga, and Tuvalu. Other island groups like Tahiti, Polynesia's largest island, are called French Polynesia because they remain under French rule.

The largest numbers of Polynesians live on the Samoan Islands. Samoans practiced **horticulture**, the raising of plants and fruits on small plots of land. Women gathered wild plants and were weavers. Most Polynesians today share similar languages and culture. Polynesians arrived in Hawaii around A.D. 500 and had profound effects on the culture.

Density and Distribution

Oceania covers a vast area. Most of the islands are not suited for human habitation. The area's population is unequally distributed among the island countries. Papua New Guinea has about 5.9 million people. The world's smallest republic, Nauru, only has about 10,000 people. Most people live on the coasts and not in the usually rugged interiors.

Oceania's population is growing at the high average rate of 2.3% because the population is young. Oceania's 25,000 islands only total about 551,000 square miles. Population density varies greatly. Papua New Guinea covers so much land it only has 33 people per square mile.

History and Government

Outside influences on indigenous cultures have shaped Oceania's societies. Migrations over generations, European colonization, and independence have shaped the nations of Oceania.

Early Migrations

Asian migrants in family groups settled along the island coasts of Oceania. They survived on fish, turtles, shrimp, breadfruit, and coconuts. They cultivated root crops like taro and yams and raised small animals like chickens and pigs.

The well-built canoes they made allowed them to travel great distances among the islands and trade gradually developed among them. To make trading easier, people on some islands used long strings of shell pieces as money.

DID YOU KNOW?

Today, on an island off the coast of Papua New Guinea called New Britain, shell money is still traded for canned goods and vegetables at markets.

European Colonization

Europeans settled Oceania in the 1800s. They set up commercial plantations to grow sugarcane, pineapple, and other tropical products that could be sold around the world.

DID YOU KNOW?

European diseases killed so many indigenous peoples the Europeans had to bring more workers in from other Pacific islands and more distant areas like South Asia.

The mix of cultures weakened indigenous societies and led to violent ethnic conflicts in countries like the Solomon Islands and Fiji. Europeans were trying to replace traditional ways of life with European beliefs and customs.

In the late 1800s and early 1900s, Britain, Germany, France, Spain, and the United States struggled for control of various Pacific Islands. These countries sought to acquire or expand influence in the region and gain new sources of raw materials.

The two world wars changed the course of history in Oceania. After World War I, many of Germany's Pacific colonies fell to Japanese control. During World War II, some Pacific Islands like Guadalcanal and Iwo Jima were sites of ferocious battles. When Japan was defeated, its South Pacific possessions like the islands of Micronesia were turned over to the United Nations as trust territories, dependent areas the United Nations placed under temporary control of a foreign country.

Since the 1970s, most of these islands, like Palau, the Marshall Islands, and the Federated States of Micronesia, have become independent countries.

Independence

Starting in the 1960s, a number of small islands in Oceania took steps toward independence. In 1962, Samoa (formerly Western Samoa) became the first Pacific island to win freedom after periods of rule by Germany and New Zealand. Today, most of the South Pacific islands enjoy some form of independent government.

In some countries of Oceania, like the Solomon Islands, traditional beliefs are starting to revive and guide decision making. People in the Solomon Islands are again valuing egalitarian relationships, subsistence agriculture (growing only enough for their own needs), and developing a strong connection to the land.

Culture and Society

Oceanic societies have been shaped by European and Asian cultural traditions and indigenous practices. Indegenous peoples still value living in harmony with their natural environment. European colonizers introduced new customs, social structures, and cultures.

Sports and Leisure

Sports and leisure reflect the region's diversity. Western style resorts attract people to beaches where they and locals enjoy the Pacific island sport of surfing. Traditional sports like outrigger canoe racing and spearfishing are also popular. In former American territories, baseball is popular. The French introduced cycling and archery to islands they controlled. Even small communities often have facilities for these and other sports like soccer, volleyball, and tennis.

Language and Religion

Before modern transportation and communications, the vast ocean separated the peoples of the South Pacific from the rest of the world. Isolated groups developed many different languages without interference from the outside world.

DID YOU KNOW?

Of the world's 3,000 languages, 1,200 are spoken in Oceania alone. Some are spoken by only a few hundred people.

European colonization brought European languages to Oceania; French is widely spoken. In many areas of Oceania, **pidgin English**, a blend of English and an indigenous language, developed to allow better communication between different groups.

Peoples of the South Pacific islands practice various forms of Christianity. These practices are often combined with traditional religious beliefs. Christianity is the most widely practiced religion in Oceania today.

On some islands, religions were introduced by immigrants many years ago. When thousands of Indians were brought to Fiji to work on plantations, they brought Hinduism with them. Over 30% of the people in Fiji practice Hinduism.

Education and Healthcare

Education quality varies across Oceania. Missionary schools in the Solomon Islands provided primary education until the 1970s. Secondary schools and universities are now common in the Solomon Islands, Fiji, and Papua New Guinea. Different education quality can be seen in different literacy rates. Fiji has the high rate of 94% literacy. Papua New Guinea has the low literacy rates of 51% for women and 63% for men.

Expanded educational opportunities for Aboriginals and new native studies programs will hopefully bridge the gap and open a path out of what Westerners define as poverty.

Healthcare is unevenly distributed across Oceania. Many Pacific Islanders suffer from poor economies and low standards of living. On remote islands, fresh food, electricity, schools, and hospitals are often inadequate. Recently, island countries have begun to improve their quality of life with international aid.

Practice

Fill in the blank with the correct word from the information in the preceding paragraphs.

1. ____________ is a system of beliefs that connect Aboriginal people to the beginning of time.

2. New Zealand and Australia have ____________ colonial influences.

3. ____________ is widely spoken in Oceania.

4. In the early 1900s, Australia and New Zealand peacefully won their independence from __________.

5. In the twentieth century, Western and Japanese powers were interested in __________ __________ of the South Pacific.

6. __________ __________ racing is a traditional sport in the South Pacific.

7. Guadalcanal and __________ __________ were the sites of ferocious battles in the Pacific during World War II.

8. __________ was the first country in the South Pacific to become independent in the modern era.

9. __________ workers were brought in to replace native workers dying of European diseases in the South Pacific.

10. Europeans established __________ plantations in the South Pacific to feed global trade and their appetites for wealth.

Economic Activities

Remote geographic locations and challenging environments influence how people earn a living in Australia and Oceania. Agriculture is the most important activity in Australia and Oceania. New industries are contributing to national economies.

Agriculture

Agriculture is the most important economic activity in the South Pacific. Australia and New Zealand export large quantities of farm products. Australia leads the world in wool production.

Less than 5% of Australians work in agriculture. Yet most of the country's vast lands are devoted to raising livestock. The dry climate forces animals to need large areas to find vegetation they can eat.

Australia's dry climate makes only 10% of its land **arable**, or suitable for growing crops. Australian farmers use irrigation, fertilizers, and modern technology to make the best use of limited agricultural land.

Over 50% of the land in New Zealand is used for agriculture. Ranchers in New Zealand, called **graziers**, raise sheep, cattle, and red deer.

New Zealand's fertile soils permit farmers to grow barley, wheat, potatoes, and fruit.

A lack of arable land in much of Oceania limits agriculture; island farmers sometimes practice subsistence farming and fishing. Some islands have rich volcanic soil and ample rainfall. The major crop is copra, dried coconut meat. Fiji exports sugarcane, copra, and sugar. Papua New Guinea supplies coffee, copra, and cacao.

Mining and Manufacturing

There is a variety of mineral deposits in parts of the South Pacific. Australia exports gold, diamonds, opals, iron ore, and bauxite. High transportation costs make mineral extraction difficult. Public debates about Aboriginal land rights limits mining in Australia. New Zealand has a large aluminum smelting industry and Papua New Guinea's rich deposits of gold and copper have only recently been exploited.

Australia and New Zealand are the South Pacific's leading manufacturers. Because agriculture is important to both countries, food processing is their leading industry. Because Australia and New Zealand are isolated geographically, to export they must import costly machinery and raw materials. Consumer product industries in both countries concentrate on making products for the home, such as appliances.

The rest of the South Pacific is less industrialized than Australia and New Zealand. Manufacturing in Oceania is limited to small-scale enterprises like apparel production.

Trade and Interdependence

Trade between Australia, New Zealand, Oceania, and other parts of the world has increased due to improvements in transportation and communications and trade agreements. The region's agricultural and mining products are its greatest sources of export income. Countries in Oceania export copra, timber, fish, vegetables, spices, and handicrafts.

Throughout the 1900s, Australia and New Zealand traded mainly with the United Kingdom and the United States. Recently, South Pacific countries have increased trade with the neighboring countries of China, Taiwan, and Japan. Australia is a member of the Asia-Pacific Economic Cooperation (APEC) forum and is pursuing trade agreements with China and the Association of Southeast Asian Nations (ASEAN).

People and Environment

Australia and Oceania have some of Earth's richest and most diverse natural resources and wildlife. Natural resources have not always been well managed.

Unique Animals

Because Australia has been isolated for so long from other landmasses, it has many unique animal species. Kangaroos, koalas, and wallabies are some of Australia's 144 species of **marsupials**, mammals whose young mature in a pouch after they are born.

The Australian island of Tasmania is home to the Tasmanian devil, a powerful meat-eating marsupial about the size of a badger.

> **DID YOU KNOW?**
>
> Australia's strangest wildlife may be the duck-billed platypus and the echidna, a spiny anteater. They are the only two mammals in the world that lay eggs.

The introduction of nonnative animals is seriously threatening these unusual species of wildlife in Australia. Hunting dogs called dingoes were brought from Asia by migrating Aboriginals. European settlers brought sheep, cattle, foxes, rabbits, and cats to Australia. Without natural predators, these animals have multiplied and taken over the habitats of the native species.

Some of Australia's native species have become extinct. At least 16 kinds of marsupials are now endangered. Efforts to restore Australia's ecological balance include using electric fencing to keep out nonnative animals, hunting and trapping programs, introducing natural predators, and creating native wildlife reserves.

New Zealand is isolated, remote, and surrounded by ocean. It is home to many unique animal species, including seabirds that multiplied without mammalian predators. The penguin is a surprising inhabitant of New Zealand.

Human settlement and other factors threaten the animal species of New Zealand. Like Australia, the most serious threat to the native animals of New Zealand comes from **introduced species** like cats, rats, and ferrets.

New Zealand is taking steps like implementing predator control techniques and establishing island sanctuaries. More than 200 of the hundreds of small islands and islets off the coast of New Zealand are managed by the national Department of Conservation as protected reserves.

Forest, Soil, and Water

Protecting forest, soil, and freshwater resources is important throughout the South Pacific. In Australia, woodlands have been cleared for farms and grazing lands. This leaves little protection from soil erosion, which is a major problem in Australia. Overgrazing arid areas and the country's worst drought in more than a century made Australia's erosion problem worse.

TIP

Reducing deforestation is essential for soil conservation.

Countries like New Zealand, Papua New Guinea, and Vanuatu with large timber reserves are developing plans to use forest resources without damaging the environment.

Australia's freshwater sources are threatened by drought, salt, irrigation, and agricultural runoff. In the Murray-Darling River basin, one of the world's largest drainage basins, using water for agriculture and the growing city population has greatly reduced river flow. Large areas of the basin are at risk from increasing soil salinity, one of Australia's most pressing environmental problems. The replacement of native vegetation with pastures and nonnative shallow-rooted crops are a major cause for increased salinity in Australia's water.

Oceania faces challenges in managing freshwater resources. Many small coral atolls and volcanic islands have limited supplies of fresh water. These water supplies are threatened by inadequate sanitation and agricultural runoff. The standard of living remains low where there is a lack of clean drinking water. It also restricts economic growth in many countries. Better management of runoff, construction of sanitation facilities, and less expensive ways of removing salt from water will help Oceania.

The region's oceans are threatened by agricultural runoff, chemical fertilizers, and organic waste. Toxic waste endangers Australia's Great Barrier Reef and other Pacific coral reefs. Coral environments are increasingly stressed by tourists, boaters, divers, oil shale mining, and increasing water temperatures.

Environmental programs that protect and monitor the condition of the Great Barrier Reef are the key to its survival. The Australian government has taken steps to protect the complex ecosystem, like restricting fishing and creating sanctuaries to preserve the unique biodiversity of the region.

Although oil drilling and mining stopped on the reef, commercial and sport fishing, spearfishing, and collecting aquarium fish and shells are regulated by the Australian government.

Nuclear Legacy

Nuclear weapon testing has had major effects on the region's environment. In the late 1940s and 1950s, the United States and other countries carried out above-ground testing of nuclear weapons in the South Pacific. The dangers of this testing were underestimated at the time.

In 1954, the United States exploded a nuclear weapon on Bikini Atoll in the Marshall Islands. The people on Bikini Atoll were moved to safety, but those downwind of the explosion on Rongelap Atoll received massive doses of radiation resulting in deaths, illnesses, and genetic abnormalities.

Even though U.S. nuclear testing stopped, the effects of radiation exposure and environmental damage have continued for generations. Atolls affected by the testing remain off-limits to human settlement. Recent studies show hope for eventual environmental recovery.

DID YOU KNOW?

In the 1990s, the United States government provided $90 million to help decontaminate Bikini Atoll and a $45 million trust fund for blast survivors and their offspring from Rongelap Atoll.

The nuclear legacy had political effects. Antinuclear activism is a major factor in regional politics. In 1985, New Zealand banned nuclear-powered ships and those with nuclear weapons from entering its waters. Because the ban is still in effect, the United States withdrew from a defense agreement with New Zealand.

Future Challenges

Environmental concerns related to atmosphere and climate changes threaten Australia, Oceania, and other world regions. In the 1970s, scientists found a hole in the **ozone** layer over Antarctica. This hole quickly grew between 1975 and 1993 to cover more than 9 million square miles. In October 2006, the hole in the ozone covered more than 10.6 million square miles. It has since stopped expanding and is showing signs of closing back up.

The loss of the protective ozone layer may cause increased rates of skin cancer and cataracts. Both conditions are caused by overexposure to the sun's ultraviolet rays. Increased solar radiation reaching Earth may contribute to global warming, the gradual rise in Earth's temperatures.

Climate and weather in the South Pacific are very sensitive to changes in the **El Niño-Southern Oscillation** (**ENSO**) weather pattern. This seasonal weather event, called El Niño, causes droughts and powerful cyclonic storms in the South Pacific. These ENSO-related weather patterns are increasing in frequency and severity and may be linked to global warming.

Continued increases in Earth's temperatures could cause the polar ice caps to melt and raise ocean waters. This would cause many of Oceania's islands to be flooded. Rising ocean temperatures could cause more of certain types of algae and plankton to grow. The overgrowth would choke out other life forms. Diatoms, plankton that grow in cold ocean waters, would die if temperatures rose. That would affect life forms that feed on them. Scientists in the region, particularly Antarctica, are studying global warming to determine causes, predict consequences, and provide solutions.

Practice

Insert T or F to indicate whether the statement is true or false, based on what you've read.

11. ____ Over Antarctica, there is a hole in the ozone layer that is more than 10 million square miles.

12. ____ In 1954, the United States exploded a nuclear weapon in the Marshall Islands.

13. ____ Native Australian marsupials like kangaroos, koalas, and wallabies have benefitted from the introduction of sheep, cats, and cows to their land.

14. ____ Oceania's freshwater supplies are threatened by inadequate sanitation and agricultural runoff.

15. ____ Levels of salinity have decreased in Australia's water as more shallow-rooted crops are introduced.

16. ____ Nonnative species of animals have taken over Australian habitats.

17. ____ The only two known species of egg-laying mammals are indigenous to Oceania.

18. ____ Maori farmers brought root crops like taro and yams to New Zealand in the 1900s.

19. ____ Australia and New Zealand have ample mineral deposits.

20. ____ Australia dedicates most of its land to livestock grazing.

Answers

1. Dreamtime
2. British
3. French
4. Great Britain
5. Raw materials
6. Outrigger canoe
7. Iwo Jima
8. Samoa
9. Asian
10. Commercial
11. T
12. T
13. F
14. T
15. F
16. T
17. F
18. F
19. T
20. T

LESSON

25

SUBCONTINENTS (PART I)

LESSON SUMMARY

This and the following two lessons focus on the physical and human geography of the **subcontinents**. The subcontinents include: Southwest Asia, including the Middle East and North Africa; Central Asia; South Asia, or Southern Eurasia; and Greenland.

We examine the physical geography—landforms, water systems, natural resources, climate, and vegetation—of the subcontinents. We also look at the human geography—population, culture, language, religion, economy, education, healthcare, arts, and family life—of the region.

Forget not that the earth delights to feel your bare feet and the winds long to play with your hair.

—Kahlil Gibran, Lebanese-American writer and mystic

Countries

Afghanistan
Algeria
Armenia
Azerbaijan
Bahrain
Cyprus
Egypt
Georgia
Iran
Iraq
Israel
Jordan
Kazakhstan
Kuwait
Kyrgyzstan
Lebanon
Libya
Morocco
Oman
Qatar
Saudi Arabia
Syria
Tajikistan
Tunisia
Turkey
Turkmenistan
United Arab Emirates
Uzbekistan
Yemen

Physical Geography

The huge region of Southwest Asia—including the Middle East and North Africa—and Central Asia, is where ancient civilizations began thousands of years ago in river valleys. Ancient rivers like the Nile remain vital to the people of the region because water is essential in an arid land. Dramatic landforms can be found Southwest Asia even though it is a region dominated by deserts and mountains.

Earthquakes

The African, Arabian, Anatolian, and Eurasian plates come together in Southwest Asia and Central Asia, and the regions regularly experience earthquakes. In 1999, Turkey, at the intersection of the Arabian and Eurasian plates, felt the effects of an earthquake measuring 7.4 out of 10 on the Richter scale. Tectonic shifts built the Atlas Mountains in Morocco and Algeria, the Zagros Mountains of southern Iran, and the Taurus Mountains of Turkey.

Mountains

The Atlas Mountains are Africa's longest mountain range. They reach across Morocco, Algeria, and Tunisia. The northern slopes have Mediterranean climate and support farming.

The Taurus Mountains are located on the southern part of Anatolia, which comprises the western two-thirds of Turkey. Enough precipitation falls on the northern side of these mountains to water the coastal regions for farming.

In Southwest Asia, the Hejaz and Asir mountains stretch along the western coast of the Arabian Peninsula. The taller Asir Mountains receive more precipitation and are thus more productive agriculturally.

The Pontic and Taurus mountians rise from the Turkish landscape. Between these ranges stands the Anatolian Plateau, 2,000 to 5,000 feet above sea level. East of the Pontic range, north of Mount Ararat between the Black Sea and Caspian Sea, are the Caucasus Mountains.

West of the Tian Shan range, which spans Kazakhstan, Kyrgyzstan, and western China, the dry Turan Lowland also has irrigated farmland. Dune-covered **kums**, or deserts, contrast with the farmlands. The Kara-Kum, or black sand desert, covers Turkmenistan. The Kyzl-Kum, or red sand desert, covers half of Uzbekistan. To the west, the Ustyurt Plateau has salt marshes, sinkholes, and caverns.

Coastal Plains, Seas, and Peninsulas

The region's agricultural base can be found along the Mediterranean Sea, the Caspian Sea, and the Persian Gulf. Southwest Asia and Central Asia include a variety of seas and peninsulas. The Mediterranean Sea separates Africa and Europe. The Red Sea and the Gulf of Aden separate the Arabian Peninsula from northeastern Africa. The Persian Gulf is to the east of the peninsula and the Arabian Sea is to the south. In the northwest, the Gulf of Suez and the Gulf of Aqaba meet the Sinai Peninsula. There are ample oil and natural gas reserves on the eastern side of the Sinai Peninsula.

Across the Mediterranean Sea in the north, the Anatolia peninsula, the westernmost tip of Asia, points west to the Aegean Sea. The Black and Mediterranean seas lie to the north and south of Anatolia. The Dardanelles, the Sea of Marmara, and the Bosporus strait connect the Black and Aegean seas and separate Asia and Europe.

East of the Mediterranean Sea are three landlocked saltwater bodies. The Dead Sea sits at the mouth of the Jordan River. In Central Asia, the Caspian Sea, touching Asia and Europe, remains the largest inland body of water on Earth.

East of the Caspian Sea is the Aral Sea, sandwiched between Kazakhstan to the north and Uzbekistan to the south. It has been reduced to a fraction of its size since the 1960s, when the Soviet Union began diverting water from the rivers that fed it for irrigation projects. By the late 1980s, the sea split into two separate bodies, the North and South Aral seas. The water levels in the North Aral Sea increased by 2006, after dams were built to ensure the flow of freshwater to the sea.

Plains and Plateaus

Early civilizations thrived in the Fertile Crescent between the Tigris and Euphrates river valleys of Southwest Asia. The ancient civilization of Mesopotamia, "land between two rivers" in ancient Greek, was irrigated by a network of canals that supported farming. Now Iraq, the area is largely arid and barren. The two rivers are thirty miles apart and they meet in southern Iraq to form the Shatt al Arab, which empties into the Persian Gulf. The Euphrates is about 1,700 miles long and the Tigris is 1,180 miles long. Dams control both rivers and hydroelectric power plants provide electricity.

Water Systems

For thousands of years, people have relied on the region's rivers and fertile river valleys where civilizations developed. Rivers nourish Southwest Asia, North Africa, and Central Asia.

The Nile River in Egypt is the world's longest river at 4,160 miles. The Nile delta and fertile land along its banks gave birth to one of the world's earliest civilizations.

DID YOU KNOW?

Nearly all of Egypt's people live on 3% of Egypt's land near the Nile delta and river banks.

Several dams control the Nile's flow, reducing flooding and the deposits of alluvial soil made of sand and mud left by moving water. In particular, the Aswan High Dam provides Egypt with water for agriculture as well as hydroelectric power. Lake Nasser, a human-made reservoir created when the dam was built, stores water and regulates the flow of the Nile. The water in the lake has brought more land under irrigation and has converted flood land to irrigated farmland.

Streambeds

Arid Southwest Asia and North Africa have streams that appear and disappear very quickly. In deserts, runoff from rare rainstorms creates **wadis**, streambeds that remain dry until a heavy rain. Irregular rainstorms often create flash floods. Wadis fill with so much sediment they become mudflows, moving masses of wet soil dangerous to humans and animals.

Natural Resources

Some of Southwest Asia, North Africa, and Central Asia have abundant resources important to the world economy.

Oil and Natural Gas

Over 63% of the world's known oil reserves and 50% of its natural gas are located in this region. Reserves under the Gaza Strip and Caspian Sea have not been measured yet.

Petroleum exports have enriched the region.

Minerals

Turkmenistan has one of the world's largest deposits of sulfur and sulfate used in paperboard, glass, and detergent. Morocco is a major producer of **phosphate**, a chemical used in fertilizers. Chromium, gold, lead, manganese, and zinc are also spread across the region. Recent discoveries indicate that the region may hold up to 10% of the world's iron ore reserves.

Large lithium reserves were discovered in Afghanistan in 2010.

Building Diverse Economies

Some countries in the region are decreasing their reliance on oil and mineral exports and diversifying their economies. The United Arab Emirates is investing oil earnings in banking, information technology, and tourism. Libya, which depends on oil for 95% of export income, is investing in agriculture, fisheries, and infrastructure.

Climate Challenges

A lack of precipitation in Southwest Asia, North Africa, and Central Asia affects the region's climates, vegetation, and human activities. Because large parts of the region receive less than 10 inches of rainfall a year, most of it contains arid areas and experiences desert climate.

Ancient cave paintings show that North Africa was once wet and green. Now, vast stretches of sand with an occasional watering hole mark the region.

Desert Climate

Deserts, which receive less than 10 inches of rainfall a year, cover almost 50% of Southwest Asia, North Africa, and Central Asia. The Sahara Desert, covering 3.5 million square miles, is the largest desert in the world and covers most of North Africa. Droughts have expanded the Sahara in recent decades. Sand covers about 20% of the Sahara, and the rest is desert pavement, or stony plains covered with gravel known as regs, barren rock, and mountains.

The desert has extreme weather patterns. The deserts of the northern parts of Central Asia have relatively cold winters with freezing temperatures. Winters in the Arabian Desert and Sahara are milder. Summers in these deserts are long and hot.

The largest sand area in the region, the Rub' al-Khali, or Empty Quarter, covers 250,000 square miles. It is one of several deserts on the Arabian Peninsula and covers most of the southern part of the peninsula.

The arid deserts in the region support vegetation like drought-resistant shrubs and cacti. In Central Asia's Kara-Kum, nomadic herds of goats, sheep, and camels graze on brush. In an **oasis**, or place where underground water surfaces, small-scale farming is possible. Villages, towns, and cities developed around numerous Saharan oases.

Steppe Climate

The second-largest climate in Southwest Asia, North Africa, and Central Asia is steppe. Steppe, which lies to the north and south of the Sahara, borders desert climates across Turkey to eastern Kazakhstan. Precipitation in this semiarid climate averages less than 14 inches a year. This supports short grasses, shrubs, and some trees which provides pasture for goats, sheep, and camels. Pastoralism is a way of life for people living on the steppe.

Midlatitude Regions

Countries in the midlatitudes benefit from rainfall in the Mediterranean, highland, and humid subtropical climates. Mediterranean climates have hot, dry summers and cool, rainy winters. This climate is common in the Tigris-Euphrates river valley, the highland areas, and on the coastal plains of the Mediterranean, Black, and Caspian seas.

Rainfall

Highland and coastal areas near mountain ranges receive the most rainfall as warm, moist air is driven off the sea by prevailing winds. The North African coast near the Atlas Mountains receives about 20 inches a year, nourishing forests. Monsoons arrive on the coast of Oman in July and August and their rains keep forests and pastures lush. Near the Elburz Mountains in Iran, more than 60 inches of rain falls a year. Batumi, Georgia, one of the region's wettest places, receives more than 100 inches of rain a year.

Exports and Tourism

Countries with Mediterranean climates like Morocco and Tunisia export olives, citrus fruits, and grapes to Europe and North America to supplement their national incomes. Tourism is also lucrative, as people from colder climates seek the sun and warmth of the Mediterranean. European tourists, for example, go to Agadir, Morocco, where the sun shines for 360 days of the year, and also visit the ancient cities of Fès, Casablanca, and Marrakech.

Areas of higher elevation like the Caucasus Mountains have a highland climate; they are usually colder and wetter than other climates in the region. The highland climate varies with elevation and exposure to wind and sun.

North Africa—Human Geography

North Africa has been home to many ethnic groups and cultures who, despite modernization and urbanization, hold on to many traditional ways of life. The Sahara and access to water have profoundly affected the people of North Africa.

Population Patterns

Indigenous ethnic groups, migrations, and the dramatic climate have shaped population patterns in North Africa. The indigenous cultures of North Africa have mixed with those of the Arabian Peninsula and Europe to form distinct cultures.

People

European immigration and colonialism have influenced the coast of North Africa—especially in Morocco, Algeria, and Tunisia—for hundreds of years. Romans, Jews, Muslims, and Spaniards have all influenced the North African subregion. Indigenous and Arab cultures dominate the rest of North Africa. Indigenous North Africans preceding the Arab invasions are called Berbers, pastoral nomads who moved based on the season and availability of grass and water for grazing. Today, most are farmers. Although estimates of Berbers range from 14 million to 25 million throughout North Africa and Europe, their numbers are highest in the Atlas Mountains and Sahara.

Arabs are the other major ethnic group in North Africa. They first left the Arabian Peninsula and came to North Africa in the A.D. 600s. Arab-speaking **Bedouins** migrated from deserts in Southwest Asia. They herd animals in the desert where there is vegetation and water to grow food on oases. Egypt was the primary gateway to North Africa.

Density and Distribution

Geographic factors, especially water availability, influenced settlement in the region. For centuries, people settled along seacoasts or river deltas because of water scarcity elsewhere.

The major urban population centers of North Africa are: Casablanca, Morocco; Algiers, Algeria; Tunis, Tunisia; Tripoli, Libya; and Cairo, Egypt. As Egypt's primate city, Cairo dominates the country's social and cultural life. People move there from the rural areas looking for a better life. Urban growth has been so quick that jobs, housing, and infrastructure like streets and utilities have lagged behind.

History and Government

Numerous migrations and invasions have influenced different cultures throughout the history of North Africa. Because the region is so close to Europe and Southwest Asia, indigenous Berbers faced migrations and invasions by Arabs and Europeans.

Early Peoples and Civilizations

About 10,000 years ago, at the end of the last Ice Age, hunters and gatherers settled North Africa. Along the Mediterranean Sea and the Nile River valley, Egyptian civilization developed about 6,000 years ago. The Nile deposited rich soil on the floodplain each year.

Egyptians used sophisticated irrigation systems to water crops in the dry season so they could grow two crops a year.

DID YOU KNOW?

The ancient Egyptians used a 365-day solar calendar, built massive temples and pyramids as tombs, and invented **hieroglyphics**.

Invasions

In the A.D. 600s, invaders from the Arabian Peninsula influenced the native Berber culture. The Berbers readily assimilated with Arab cultures in Morocco and Algeria and did so to a lesser extent in Tunisia where, following Vandal and Byzantine invasions, Arab rule was established in Tunisia. Arab and Turkish culture dominated North Africa until the end of the Ottoman Empire in 1922.

Internal Arab invasions from the east brought the religion and culture of Islam to Morocco. Muslim and Jewish exiles fleeing persecution in Catholic Spain brought Spanish culture to Morocco in the 1400s. In Algeria in the 1500s, early Berber-Arab rule was overthrown by the Ottoman Empire.

DID YOU KNOW?

The Ottoman Turks also dominated the Mediterranean Sea for 400 years, forcing Europeans to seek other land and sea routes to trade in the East.

European Colonialism

As Europe became more skilled at sea travel, the Arab and Ottoman Turkish empires grew weaker. European colonial rule influenced the people and culture of North Africa. In the mid-1800s, Algeria's independence was cut short by French invasion and conquest. The European colonial empires drew linear geometric boundaries for Libya, Egypt, and Algeria without considering natural or cultural characteristics. Because local governmental practices differed from imperial designs, there was conflict.

In the 1800s, a well-educated urban middle class developed in North Africa. Trained in the European idea of nationalism, North Africans learned that ethnic groups have the right to their own countries.

Independence

Egypt became independent of the United Kingdom in 1922. Trade between the Mediterranean Sea and the Red Sea through the Suez Canal made Egypt a regional power. Egypt is also a center of Arab nationalism.

In the mid-1900s, Algeria's strong nationalist movement achieved independence through civil war. Since independence in 1962, Algeria has developed its resources and raised its standard of living.

In the 1950s and 1960s, other countries in North Africa achieved independence from European colonialism. Libya won independence from Italy in 1951, but was still ruled by a strong pro-Western monarchy. Until 1969, when the monarchy was overthrown in a coup led by Colonel Muammar al-Qaddafi. Qaddafi ruled oil-rich Libya until civil war broke out in 2011. Tunisia achieved independence from France in 1956. Morocco won independence from France the same year and established a constitutional monarchy that still rules the country today.

Religion and Language

Most people in North Africa are Sunni Muslims. Sunnis believe leadership in Islam belongs to the community. Shias believe leadership in Islam belongs to the descendants of Muhammad's cousin Ali. Most Berbers have accepted Islam, but some practice indigenous religion.

In countries with large Muslim populations, calls to prayer occur five times a day. The *muezzin*, or crier, calls the faithful from the local mosque's minaret or tower. Following the movements of the imam, or prayer leader, worshippers bow, kneel, and touch their foreheads to ground in the direction of the holy city of Makkah, or Mecca, in Saudi Arabia.

The Arabic language spread with Islam across North Africa. Non-Arabic Muslims learned Arabic to read Islam's holy book, the Quran (or Koran). As more people joined Islam, Arabic became the region's primary language.

Education and Healthcare

Most young people in North Africa go to school. Primary education is free and increasing. Literacy rates vary. Morocco's is 52%. Libya's is 82%. Healthcare has improved in the last few decades. People usually go to government-owned hospitals for medical treatment. Due to doctor shortages, there is limited health care for rural people.

Arts

The ancient Egyptian pyramids are an engineering wonder set to coordinate with star constellations. The Egyptians combined the sciences of mathematics and astronomy with spirituality.

The art of weaving, embroidering, and metalworking was highly influenced by Islam. Because many Muslims believe that creating a human form is sinful, Islamic arts tend to contain intricate patterns rather than pictures. A popular form of music in Algeria, *rai*, features a number of instruments and poetic lyrics. Whirling dervishes dance in a Sufi mystic Muslim trance.

Eastern Mediterranean—Human Geography

Ancient civilizations and cultures continue to influence the Eastern Mediterranean subregion. This region is where the religions of Judaism, Christianity, and Islam were born. Palestinian and Jewish people claim rights to the land of Israel.

Population Patterns

Migrations, claims to ancestral homes, and boundary disputes have influenced population in the Eastern Mediterranean. About 7.5 million people live in Israel. Almost 80% are Jews who trace their heritage back to the Israelites who settled in Canaan, or modern-day Israel, and Lebanon.

Density and Distribution

To avoid the desert climate, most people in the Eastern Mediterranean live along the coastal plains or the Euphrates river valley. Because of a lack of water in many areas, small land areas, and large human populations, the area has some of the highest population densities in Southwest Asia; Lebanon, for example, has 1,046 people per square mile.

The Eastern Mediterranean subregion is mostly urban. More than 92% of the Israeli population lives in cities. Many people live in Tel Aviv-Jaffa and other coastal cities of central Israel. Over 50% of Syrians and Palestinians live in urban areas. A majority of Lebanon's population lives in coastal cities like Beirut and Tripoli.

Modern Israel has faced intense immigration since its founding in 1948. Since 1989, about a million Jews migrated from the former Soviet Union. Because people have migrated to Israel from at least 100 countries, the country remains ethnically diverse.

History and Government

The eastern Mediterranean is home to three of the world's major religions that have shaped politics and culture for centuries. Jerusalem is the Jewish capital and religious center kings David and Solomon built. The city is also sacred to Christians because many of Jesus' acts took place there. Mohammed also ascended to heaven in Jerusalem.

Early Civilizations

Over the centuries, the Eastern Mediterranean has been occupied by powerful cultures and empires. The Phoenicians widely traveled the Mediterranean starting about 3000 B.C. They created an alphabet with letters representing sounds, an idea that influenced many modern alphabets. Ebla, Syria, was an important commercial center by 3000 B.C. By 2400 B.C., hundreds of thousands of people lived in this Semitic empire. Damascus, Syria, remains one of the oldest, continuously settled cities in the world.

Religion

Judaism, Christianity, and Islam share many beliefs such as **monotheism**, or belief in one God; life after death; and a claim to Jerusalem.

Independence and Conflict

For centuries, Islamic empires called caliphates rose and fell in the Eastern Mediterranean. Physical geography restricted economic development. The region's Islamic empires lacked resources like minerals, wood, and coal used in the Industrial Revolution of Western Europe. Western European powers controlled large parts of the Eastern Mediterranean by the late 1800s.

Eastern Mediterranean countries achieved independence around the time of World War II. Lebanon gained independence from France in 1943. In 1946, Syria became independent of France. The same year, Jordan gained independence from the British.

Modern Arab-Israeli Conflict

After the Romans drove the Jews from their homeland in A.D. 135, they settled in communities across the world where they continued to face persecution. In the late 1800s, Zionists began calling for return to and resettlement of Palestine the homeland.

The British promise of an Arab homeland in Palestine convinced many Arabs to fight for the British in World War I. The United Nations divided Palestine into a Jewish state and an Arab state in 1947. When British forces left the region the next year, Jews proclaimed the independent state of Israel. Disputes over land ownership and settlement led to six wars. In 1948 and 1967, Israeli forces occupied Arab lands.

Israeli-Palestinian Peace Process

Many Palestinians became refugees in other lands. The status of Palestinian refugees and the need for a stable Palestinian state remain important issues in the Arab-Israeli conflict.

Israeli and Palestinian leaders made strides toward peace in the 1990s, but the peace process stalled in 2002. There have been numerous steps forward and backward since.

Israel has created a wall to separate the West Bank from Jerusalem. Palestinian terrorists often come from refugee camps in Palestine. The United States, European Union, United Nations, and Russia have all called for a "Road Map to Peace," to include establishment of a Palestinian state. Israel claims the entire city of Jerusalem; the Palestinians want east Jerusalem to serve as capital of the Palestinian state.

Religion and Language

Most Muslims in the eastern Mediterranean are Sunni. Arabs in Syria and Lebanon are Shia.

Jews and Christians are only a small part of the population of the Eastern Mediterranean. Most Jews live in Israel. Large groups of Christians live in Syria and Lebanon, where they remain a minority.

Family life in the Eastern Mediterranean often involves extended family and religious worship. During the holy month of Ramadan, Muslims fast during daylight.

Western-style dress is gaining popularity, particularly among Jewish and Christian women of the eastern Mediterranean. Many Muslims dress conservatively. Some Muslim women wear a veil over their faces and completely cover their feet and hands.

Arabic is the primary language of the Eastern Mediterranean. Hebrew is spoken in Israel. Hebrew and Arabic are Semitic languages. English is spoken in some areas.

Education and Healthcare

Education is compulsory and free in the Eastern Mediterranean. Most young people attend school. Literacy rates vary from 97% in Israel to 80% in Syria. Healthcare has improved over the past few decades. Many hospitals are government-owned.

Arts

For thousands of years, the people of the Eastern Mediterranean have expressed themselves through art and architecture. The Phoenicians who lived along the coast created the first alphabet. Jewish, Christian, and Muslim artists and writers have been inspired by the religions of the region.

Muslim scholars wrote about Islamic achievement and translated classic Greek texts into Arabic. These works added to the European knowledge of the ancient world. Syrian arts and scholarly writings rivaled those of Mesopotamia and even influenced Roman culture and thought.

Practice

Fill in the blank with the correct word from the information in the preceding paragraphs.

1. ___________ is a semiarid region.

2. The ___________ is the largest desert in the world.

3. Southwest Asia has a large reserve of ___________ and ___________.

4. The ___________ High Dam in Egypt reduces flooding of the Nile.

5. ___________ is a scarce resource in Southwest Asia.

Answers

1. Steppe
2. Sahara
3. Petroleum and natural gas
4. Aswan
5. Water

LESSON

26

SUBCONTINENTS (PART II)

LESSON SUMMARY

Discussion of the Subcontinents continues with the northeastern Mediterranean, Central Asia, and Greenland.

Afghan society is very complex, and Afghanistan has a very complex culture. Part of the reason it has remained unknown is because of this complexity.

—Mohsen Makhmalbaf

Northeast Mediterranean

Descendants of ancient civilizations affect society, politics, and culture today.

Population Patterns

Ethnic diversity and Islam have profoundly shaped the northeastern Mediterranean. Muslim religious traditions have also shaped the subregion.

People

People came to the northeastern Mediterranean from Central Asia and the Arabian Peninsula. Despite diverse languages and customs, Islam connects the people of the subregion.

Turks

For 8,000 years, many peoples occupied Anatolia, or modern-day Turkey. Turkic people arrived on the peninsula from Central Asia around A.D. 1000. The Ottoman Turks ruled the eastern Mediterranean for 600 years. Turks practice Islam and their culture blends Turkish, Muslim, and Western elements.

Iranians

Iran was called Persia until 1935. Iran means "land of the Aryans." Many of the 67 million Iranians believe they are descendants of the Aryans, Indo-Europeans who came from southern Russia about 1000 B.C. Persians, or Iranians, retained their culture and language, Farsi, despite numerous invasions. About 89% of Iranians are Shia Muslims.

Arabs

With ethnic ties to the Arabian Peninsula, most people in Iraq are Arab. Most Arabs in Iraq are Shia Muslims. About 35% of Iraq is Sunni. Saddam Hussein was a Sunni Muslim. Arabic is the language of Iraq.

Kurds

Kurds have lived in the mountainous areas of Turkey, Iran, and Iraq. Most Kurds speak Kurdish, which is related to Farsi, and practice Sunni Islam. Kurdish customs and dress differ from those of Arabs in the region. They have no country of their own, but call their region Kurdistan. Their efforts at self-rule have been crushed by Turkish and Arab rulers.

Density and Distribution

Iran and Turkey have populations of about 70 million each. Iraq has a population of about 29 million. More than half the people live in cities. Most Iraqis live in a strip from Baghdad to the Persian Gulf. Regional population density ranges from 110 people per square mile in Iran to 244 people per square mile in Turkey.

The large urban centers of Tehran, Iran; Istanbul, Turkey; and Baghdad, Iraq, influence social and cultural life in their countries. Cities are overcrowded. Traffic control systems and public transportation have expanded. Iran has moved some of its governmental offices to towns and villages in an effort to expand services and slow urban growth.

History and Government

Ancient empires and thriving civilizations influenced the early history of northeast Mediterranean. Today the subregion is shaped by the oil industry and relations with the outside world.

Civilizations and Empires

The area between the Tigris and Euphrates River where Mesopotamia developed was one of the world's first influential cultural hearths. The rich agricultural region was known as the Fertile Crescent.

Sumerians used irrigation canals to grow crops year-round. Sumerians created a code of laws and made advancements in mathematics, astronomy, and engineering; they kept records using a wedge-shaped writing system on clay called **cuneiform**.

The Persian Empire stretched across the region in the 500s B.C. Persian engineers designed underground canals called *qanats* to overcome evaporation and deliver water from mountains to farmlands.

The Ottoman Empire that lasted for 600 years (1299–1923) was based in modern-day Turkey. It covered North Africa, Western Asia, and Southeastern Europe.

Modern Era

By the late 1800s, Western Europe controlled northeastern Eurasia. Britain controlled Iraq until 1932. After that, Iraq suffered from turmoil, including the U.S. wars against Iraq in 1991 and from 2003 to 2011.

The Ottoman Empire was destroyed by World War I. Turkey established itself as a secular republic in 1923. It is poised to join the European Union.

The Zagros Mountains form a natural barrier between Iraq and Iran. In the Islamic Revolution of 1979, radical and fundamentalist Muslims in Iran overthrew the U.S.-supported shah. Muslim clerics called mullahs came to power and radicals still dominate Iranian politics despite civil unrest in urban centers.

Oil

The discovery of oil in the Persian Gulf in the early 1900s gradually brought riches to the region. The wells fell to regional powers. The Gulf States of Iraq, Iran, Saudi Arabia, and Kuwait joined Venezuela to form the Organization of Petroleum Exporting Countries (OPEC). They agreed on regulating production to keep oil prices high. With growing demand, OPEC grew in power.

Language and Religion

Although most Muslims speak Arabic, Iranians speak Farsi and people in Turkey speak Turkish. Most people in Iraq and Iran are Shia Muslims. Most Muslims in Turkey are Sunni.

Education and Healthcare

Education is required to grade 6 in Iraq and to grade 8 in Turkey. The literacy rate is 77% in Iran and 87% in Turkey. Before the revolution in Iran, less than half the people could read or write. Since the revolution, there have been reforms diminishing secular ideas and emphasizing religious ones.

Healthcare varies in the subregion. Since the Persian Gulf War of 1991 when Iraq attacked Kuwait, Iraq has struggled to rebuild its hospitals. In other countries, government-run hospitals suffer from doctor shortages in rural areas.

Arts

Early civilizations created large buildings, fine metalwork, and sculptures. The Sumerians built large mud-brick "stairway to heaven" temples called ziggurats. They were pyramid-shaped and stood straight up from the flat landscape. Literature from the northeastern Mediterranean is based on oral traditions, poetry, and epics. The *Rubáiyát* by the Persian poet Omar Khayyám comes from this tradition.

Arabian Peninsula

The harsh desert climate and coastal regions, as well as Islam, have shaped the Arabian Peninsula. The nomads of the subregion have adapted to the climate, mixing tradition and religion with commerce.

Population Patterns

A shared religion, a common language, and rapid modernization have formed today's Arabian Peninsula.

People

Most of the 56 million people of the Arabian Peninsula live along the coast. Most are Arab and Muslim. Islamic religion and the Arabic language are still influential here.

Arab people lived here before the rise of Islam. Many people here descend from Egyptians, Phoenicians, Saharans, Berbers, and other Semitic speakers.

DID YOU KNOW?

Many Arabs migrated to Kuwait when oil was discovered there in the 1900s.

Many South Asians also live in the larger cities of the eastern Arabian Peninsula. Muslims from

India, Pakistan, India, Bangladesh, and Iran have also immigrated here looking for jobs.

Density and Distribution

As in North Africa, the region's harsh desert climate pushed people to the lush coastal climates of the Arabian Peninsula. Some Bedouin nomads roam the vast Arabian Desert, but many have migrated to cities. The population density of some cities and oases in Saudi Arabia is 2,600 people per square mile, but the country's immense size and harsh desert climate makes its total population density average to only about 30 people per square mile. In Bahrain, 89% of people live in the cities of Manama and Al Muharraq. In Oman, more than half the population lives along the coastal plain.

Arab people in Yemen are not nomads. They settled in villages and small towns. Bedouin nomads have historically roamed the desert looking for water and grazing spots for their herds of cattle.

DID YOU KNOW?

Over 95% of the Bedouin population of Saudi Arabia is now settled.

Since the discovery of oil in the early 1900s, Arab countries have experienced increased wealth, modernization, and migration. Newly arrived immigrants make up most of the population of many countries like the United Arab Emirates; only 20% of its people are citizens. Foreign workers make up 50% of the population in Qatar and 60% of the population in Kuwait.

History and Government

Conquering empires and unified governments have imposed cultures on the peoples of the Arabian Peninsula that are still influential today. Many of the countries are young, but their cultural history is long. In the 1800s, the people sought British protection against invaders like the Ottoman Turks.

Early Cultures and Conquests

Vital cultures have existed on the Arabian Peninsula for 5,000 years. One center of civilization, Yemen, thrived from 1100 B.C. to the A.D. 500s.

The harsh desert climate of the peninsula made the coasts the only place large settlements could exist. An island near Kuwait was used as a trading post by the ancient Greeks for 2,000 years.

A local family took control of the Arabian Peninsula in 1750. The native people struggled against invading Ottoman Turks and others. The United Kingdom of Saudi Arabia was established in 1932.

In the 1800s and early 1900s, Kuwait, Bahrain, and Qatar allied with Great Britain to protect themselves from the Ottoman Empire. Parts of Yemen also allied with the British during that time. Oman has remained independent.

Language and Religion

Makkah, or Mecca, in Saudi Arabia is the sacred Muslim pilgrimage site. Most people on the Arabian Peninsula are Sunni or Shia Muslims. Most Muslims believe in making a pilgrimage, or **hajj**, to Makkah at least once during their lives.

A different form of Islam, **Ibadhism**, is practiced in Oman. This form of the religion exhibits a moderate conservatism by choosing its ruler by communal consensus and consent. Oman is the only Muslim country to have a majority Ibadhi population.

The Wahabi **sect** of the Sunni branch originating in Saudi Arabia advocates the literal interpretation of the Quran and independence from Western imperialism.

Islam influences the language in the Arabian Peninsula—the Quran is written in Arabic and most people speak and pray in Arabic. South Asian and Afro-Asian languages are found on the Arabian Peninsula, too.

Education and Healthcare

Most children in the subregion attend school. Public education is important to Kuwait and Qatar. Kuwait has a literacy rate of 93% and in Qatar it is 89%. Bahrain uses its oil revenues for education, so its literacy rate is 87%. Oman has made secondary and post-secondary education a priority because the government wants a trained workforce.

Healthcare varies by country. Rural and urban healthcare services vary. Government-owned hospitals are often understaffed. Private social or religious groups provide healthcare and social services.

Arts and Celebrations

Some of the best art on the Arabian Peninsula is found in architecture. **Mosques** and palaces are fine examples of Islamic architecture. Because Islam condemns depicting living figures in religious art, Muslim artists work in geometric and floral designs. They also use artistic writing, or calligraphy, for decoration. Passages from the Quran adorn the walls of many mosques.

Muslim families and community (*umma*) come together for religious holidays and observances. Many Muslims mark Id al Adha, the Feast of Sacrifice, by making a pilgrimage to Makkah. They also observe the holy month of Ramadan by fasting from dawn until dusk.

Practice

Insert T or F to indicate whether the statement is true or false, based on what you've read.

1. _____ The United States armed the mujahedeen of Afghanistan in their struggle against the Soviet military.

2. _____ Islam was born on the Arabian Peninsula.

3. _____ The Quran is written in Farsi.

4. _____ Hebrew and Arabic are Semitic languages.

5. _____ Oil has meant urban population growth for Southwest Asia.

Central Asia

Population Patterns

Invasions, domination by many empires, and rugged landscape have created challenges for Central Asia. Some people were defeated by war or famine or absorbed by more powerful groups. Others have survived and flourished.

People

Afghanistan sits at the southern end of Central Asia south of Turkmenistan, Tajikistan, and Uzbekistan. Afghanistan's many ethnic groups reflect centuries of migrations and invasions. Pashtun is the major ethnic group.

Over 50 ethnic groups and nationalities live in the Caucasus area. Armenians and Georgians are the most numerous. Armenia and Georgia became independent when the Soviet Union dissolved in 1991.

Many Turks live in the Central Asian republics. Uzbeks and Kazakhs are large Turkish ethnic groups. They make up the populations of Uzbekistan and Kazakhistan. Kazakhs make up only half the population in their own country.

Density and Distribution

The population of Central Asia is spread unevenly across the mountainous terrain. Russia is the most populous country in the subregion, with 140 million people, though because it spans from Europe to the Pacific, Afghanistan, with 29 million people, is the region's true population leader.

The population of Central Asia has often faced conflict. In 1915, between 1 million and 1.5 million

Armenians in Turkey were massacred, deported, or died by the hand of the Ottoman Turks. In the modern era, about 20% of the population left the country looking for a better life. About 300,000 people in Georgia are displaced.

Most of Tajikistan's people live in the Amu Dar'ya and Syr Dar'ya river valleys. The rivers flow through the country.

History and Government

Central Asia's location has left the region's people vulnerable to invasion for centuries and new challenges in the modern era. Newly independent, these countries work to find political and economic stability.

Cultures and Conquests

Central Asia is a crossroads of cultures at least 2,500 years old. Because Afghanistan is situated on trade routes between the Middle East and South Asia, it has often been invaded. The mountainous areas that extend into modern Pakistan were so treacherous that Alexander the Great and the British gave up trying to control the country. For the past 30 years the country's stability has been plagued by invasions and ethnic conflict.

The Kingdom of Urartu's rule over the Caucasus dates back to the thirteenth century B.C., and peaked around 800 B.C. before declining in the sixth century. The area was later occupied by the Roman Empire and adopted Western philosophical, political, and religious traditions.

Starting about 100 B.C., parts of Central Asia prospered using the Silk Road, a trade route connecting the Mediterranen Sea to China. Many cities in Central Asia, such as Samarqand in modern-day Uzbekistan, thrived as trading centers along the road.

Accessible because of its location along trade routes, Central Asia fell under the control of empires. In the A.D. 1200s, Genghis Khan's nomadic Mongol tribes from north of China invaded the region, killing tens of thousands of people. The Mongols brought improvements like paper money and safer trade routes. Alexander the Great, the Persians, Arabs, and Ottoman Turks all unified the region at different times.

Armenians resisted conquest. The country is located next to the Muslim countries of Turkey and Azerbaijan. Over 90% of Armenians practice Christianity and it is often called the first Christian country. It is an **enclave**, an area culturally or ethnically different from its surrounding cultures. Even though Azerbaijan is surrounded by the Christian Caucasus region, it also maintains an Islamic culture.

The Russian Empire unified parts of Central Asia in the 1800s. By 1936, most of Central Asia was an extension of the Soviet Union. The region was controlled politically, economically, and culturally by the Soviets, and many countries had increased literacy rates and standards of living under the Soviets. Many people left Kyrgyzstan to escape harsh living conditions.

Independence

When the Soviet Union dissolved in 1991, numerous Central Asian countries declared independence. Armenia has had some stability and economic reform. After the breakdown of central authority, Tajikistan experienced chaos and still has a Russian military presence.

The Soviet Union invaded Afghanistan in 1979 to spread communism. The United States, England, and China armed the country's guerilla rebels, the **mujahedeen**, to stop Soviet expansion. The Soviets withdrew in 1988. The mujahedeen stormed the capital of Kabul, overthrew the government, and established the Islamic Republic of Afghanistan.

A militant **fundamentalist Islamic** group, the Taliban, violently cracked down on crime, drug trafficking, and women's rights in an effort to restore order through theocratic means.

After the terrorist attacks on the Pentagon and the World Trade Center on September 11, 2001, a Saudi, Osama bin Laden, claimed responsibility and was reputed to be hiding in Afghanistan. The Taliban refused to turn him over, and the United States attacked Afghanistan in October of 2001. In 2004, Afghanistan held its first successful election, and Hamid Karzai was elected president. The Taliban has regrouped and revived as a strong insurgency movement. American forces are attempting to stabilize the region.

Afghanistan is plagued with a decentralized power structure run by warlords, a history of weak central government, corruption, and terrorism. World peace organizations and foreign aid will help build infrastructure and communications systems.

Although some Central Asian countries are moving toward political and economic stability, unemployment and poverty are widespread. Armenia, Georgia, and Kazakhistan hope oil and gas reserves will make their countries stable.

Language and Religion

The overwhelming majority of people speak a Turkic language. The Armenian, Tajikistan, Afghan Persian, and Pashto languages are Indo-European languages spoken in Armenia, Tajikistan, and Afghanistan. Georgia has a unique alphabet and language. Russian is the official language of Kazakhistan and is spoken widely in Turkmenistan and Uzbekistan.

Islam is the dominant religion, and most Muslims in Central Asia are Sunni. Most Azerbaijani Muslims are Shia. Christianity is the main religion of Armenia and Georgia. About 90% of Armenians belong to the Armenian Apostolic Church which dates back to the A.D. 300s. Most Georgians belong to the Georgian Orthodox Church.

Education and Healthcare

Education is universal throughout Central Asia and mandatory through secondary school in a few countries. Most Central Asian countries have a literacy rate above 99%. Afghanistan, at 28%, lags far behind the average.

Since the breakup of the Soviet Union, healthcare resources have gone lacking in Central Asia. Years of civil strife and economic challenges leave scant financial resources for social programs.

Central Asia Today

Economic Activities

The large oil and natural gas reserves of Central Asia have brought economic growth to some of the region's countries and have affected its relations with other regions. Most countries rich in oil have little water.

Agriculture and Fishing

Only a small part of Central Asia is suitable for farming, but a large percentage of people work in agriculture. Only about 12% of the land in Afghanistan is arable, yet 67% of the people farm for a living. Georgia's humid subtropical climate is good for growing fruits, grapes, and cotton.

The steppes of Central Asia provide fertile soil for growing crops and grasslands for grazing livestock. Uzbekistan is one of the world's leading cotton producers. Uzbekistan and Turkmenistan raise silkworms. Although only 21% of Azerbaijan's land is arable, Kazakhstan is a major grain producer.

Industry

During the 1990s, Russia underwent significant change torward a market-based economy, and by 2009 was the world's top exporter of natural gas, the second-leading exporter of oil, and the third-largest exporter of aluminum and steel.

Aside from Russia, Kazakstahn is the biggest former Soviet republic, and is one of the top twenty oil-exporting nations in the world. It also produces metals including manganese, lead, zinc, and copper, and has a large agricultural production of spring wheat. Azerbaijan's main industry is also oil and petrolem products, along with natural gas.

The region's other countries are poor and tend to rely on agriculture. During the Soviet era, Armenia modernized and began exporting textiles, machine tools, jewelry, and other manufactured goods, but since independence it has relied on farming products like livestock, vegetables, and fruits, particularly grapes. Georgia, like Armenia, has to import most of its energy needs. It produces manganese and copper. The mountainous country of Kyrgyzstan exports small amounts of gold, uranium, mercury, and natural gas, and desert-dominated Turkmenistan also exports a significant amount of gas while relying on cotton production. Tajikstan is one of the region's poorest countries, with 60% unemployment. Most of the country's laborers work abroad and send money home.

Managing Resources

Growing populations throughout North Africa, Southwest Asia, and Central Asia severely strain scarce water resources. According to the World Health Organization and UNICEF, two out of three of the world's households do not have a source of fresh water close to them.

Water Resources

Most freshwater in Central Asia comes from aquifers, underground layers of porous rock, gravel, or sand containing water. Population growth requires more water.

Troubled Seas

The Aral Sea and Caspian Sea face environmental difficulties. Pollution and overfishing threaten the southern end of the Caspian Sea near the Elburz Mountains of Iran. The Aral Sea was diverted from feeder rivers to irrigate croplands. This killed the fishing industries and dust storms spread polluted soil. People near the Aral Sea are creating a chain of lakes to support fishes.

Nuclear and Chemical Dangers

Central Asia inherited Soviet-era environmental problems. Kazakhstan had Soviet nuclear bases and nuclear, chemical, and biological weapons were tested there during the Cold War. Residents were not warned or evacuated before testing, and there were radiation leaks. Kazakhstan was used by Soviet planners for heavy industry. Industrial pollution there raised the infant mortality rate.

Practice

Insert T or F to indicate whether the statement is true or false, based on what you've read.

6. ____ Russia is the world's top exporter of natural gas.

7. ____ Afghanistan has a literacy rate on par with the rest of the region.

8. ____ Most Muslims in Central Asia are Sunni.

Greenland

The term subcontinent in the sense of "a large landmass that is smaller than any of the usually recognized continents" can also refer to Greenland. Greenland is a large island landmass, smaller than the recognized continents.

Physical Geography

Greenland is a continental island. It is part of the continental shelf, a shallow underwater platform forming a continental border. Greenland is near Ellesmere Island, which is also a continental island. Greenland is the world's largest island at 839,399 square miles.

Along the coasts of Greenland, there is sparse tundra vegetation like sedge, cotton grass, and lichens. Greenland's ice-free areas have few trees. Dwarfed birch, willow, and alder scrubs survive. As in other northern areas, few people inhabit Greenland due to its harsh climate.

Greenland's interior has an ice cap climate. This type of climate has layers of ice and snow often more than 2 miles thick constantly covering the ground.

Human Geography

Norseman, or Viking, Erik the Red arrived in Qaqortoq in A.D. 982. He was on the lam (a Viking word) after killing a man who did not return his property. Erik came from Iceland and gave the land the name Greenland to encourage settlement. It worked. Over 4,000 Vikings settled there. Vikings farmed, raised sheep and cattle, and built churches on the southern and western sides of Greenland. They traded seal and walrus skins for timber and iron from Europe.

The Inuits came from Canada around the same time. Kalaallisut, an Inuit dialect, is the official language of Greenland. About a quarter of Greenland's people live in its capital, Nuuk, which was settled by a Lutheran missionary in 1721. Greenland has been an overseas territory of Denmark's since 1721, and since 1979 has been self-governing.

Greenland Today

Greenland is melting due to global warming. Satellite imaging and ice penetrating radar shows Greenland's melting ice caps are raising the level of the oceans. Greenland is returning to the warm medieval climates that brought the Vikings from Scandinavia. Much of the ice in many areas is now too thin for sledding.

Most of the population's 56,000 live on Greenland's western coast. Fishing run by a state-owned company brings in more than 70% of the national income. Greenland imports most of its produce.

Oil beyond the fishing grounds may lead to economic self-sufficiency, but it would destroy the Inuit and Viking ways of life.

An Australian company has discovered in Greenland rare earth metals that, if mined, could be used in green technologies like hybrid car batteries, wind turbines, and compact fluorescent light bulbs. China currently dominates 95% of this market.

Denmark pumps about $620 million into Greenland's anemic economy every year. Many of the people were brought into the city by the Danish government in the 1950s and 1960s. In 2008, the people of Greenland voted for greater independence from Denmark.

Practice

Insert T or F to indicate whether the statement is true or false, based on what you've read.

9. ____ Greenland is the world's largest island.

10. ____ Greenland has Viking, Inuit, and Danish cultural influences.

11. ____ Traditional hunting and fishing has declined in Greenland due to global warming, but oil and mining are prospects for economic revival.

Answers

1. T
2. T
3. F
4. T
5. T
6. T
7. F
8. T
9. T
10. T
11. T

LESSON

27

SUBCONTINENTS (PART III)

LESSON SUMMARY

This discussion of the subcontinents ends with coverage of Bangladesh, Bhutan, India, Maldives, Nepal, Pakistan, and Sri Lanka. Due to the amount of information in this lesson, break your study up into two separate days in order to fully process what you learn.

Europe is merely powerful; India is beautiful.

—Savitri Devi

Physical Geography

The phrase *the subcontinent*, used on its own in English, commonly refers to the East Subcontinent. The region largely comprises a peninsula of Eurasia south of the Himalayas. Bangladesh, Bhutan, India, Maldives, Nepal, Pakistan, and Sri Lanka are all separated from the rest of Asia by the Himalayas but are still joined to it and thus are part of the South Asian subcontinent. South Asia contains desert, plateaus, rain forests, mountains, and a myriad of languages, races, and religions.

Landforms

South Asia's landforms affect where people live and seasonal rain patterns.

Northern Landforms

The towering Himalayas were created by the collision of tectonic plates. According to the theory of continental drift, 200 million years ago the Indian subcontinent broke off from Africa and collided with the southern edge of Asia to create the Himalayan mountain ranges along the northern edge of the peninsula. The Himalayas are about 1,500 miles long and hundreds of miles wide.

The Himalayas meet the Karakoram Mountains in the north and the Hindu Kush range in the west, creating a wall between the subcontinent and the rest of Asia. Invaders could only use the Kyber Pass between Pakistan and Afghanistan.

The region's great rivers, the Indus, Ganges, and Brahmaputra, are at the foot of the Himalayas. They water the wide fertile plains. One-tenth of the world's people live on the northern plain, the Gangetic Plain. The Chota Nagpur Plateau of forests is located in northeastern India.

Central and Southern Landforms

The Vindhya and Satpura mountain ranges across central India were also created by the collision of tectonic plates. These ranges separate the cultures of northern and southern India. The Narmada River flows through the valley between the two ranges.

At the base of the subcontinent is a triangle of rugged hills called the Eastern Ghats and the Western Ghats (Ghats means "benevolent mountains" in the ancient Sanskrit language of India). Between them is the Deccan Plateau, which was connected to Africa more than 200 millions years ago. The plateau was covered in lava that is now rich, black soil. The Western Ghats range absorbs most of the rain-bearing winds. The Karnataka Plateau, southwest of the Deccan Plateau, receive a lot of rain and is lush and green.

The island of Sri Lanka broke from the original Indian landmass. It is a chain of coral atolls and volcanic outcroppings. The Maldives cover 35,200 square ocean miles, but its land area is only 116 square miles.

Water Systems

From the north, South Asia's major river systems bring fertile soil to the floodplains during the monsoon seasons from high atop the Himalayas. These northern rivers serve as transportation routes and provide hydroelectricity.

Ganges and Brahmaputra Rivers

The Ganges River flows east from the Himalayas and empties into the Bay of Bengal. Its drainage basin covers about 400,000 miles.

> **DID YOU KNOW?**
>
> The river is named for the sacred Hindu goddess Ganga. The Ganges flows even during the dry season, December to June.

During the summer monsoon season, there can be devastating floods.

The Gangetic Plain has been cleared of grassland and forest for rice, sugarcane, and jute. It is the world's longest **alluvial plain**, or area of fertile soil deposited by river floodwaters. It is India's most agriculturally productive and densely populated area.

The Brahmaputra River flows east to the Himalayas, west into India, and south into Bangladesh. The Brahmaputra joins the Ganges River to form a delta that empties into the Bay of Bengal. The Brahmaputra is a main inland waterway. Boats can travel 800 miles into the interior using the Brahmaputra River. The Brahmaputra also provides hydroelectric power.

Indus River

The Indus River is known as the cradle of ancient India even though it flows mainly through Pakistan, watering peach and apple orchards and then emptying into the Arabian Sea. The Indus is still an important transportation route.

Natural Resources

South Asia has many natural resources on which the large population depends. Rivers are the most important resource providing drinking water, fish, alluvial soil, hydroelectric power, and transportation.

Water resource management is difficult in South Asia because rivers cross national boundaries. Bhutan works with India on hydroelectric power and they share generated power. Massive dam projects, like the Tarbela Dam in Pakistan, threaten to flood existing settlements or are choked up with built-up silt.

South Asia has large and diverse mineral stocks. India is a leading exporter of iron ore and mica. Chromium and gypsum need development. Nepal produces mica and copper. Sri Lanka is one of the world's leading makers of graphite, the lead used for pencils, and it has precious and semiprecious stones.

There are petroleum reserves near the Ganges Delta, along India's northwest coast, in the north of Pakistan, and in the Arabian Sea. There are natural gas fields in the Ganges Delta, southern Pakistan, and in Bangladesh. India has uranium north of the Eastern Ghats.

Timber resources are Indian sandalwood, teak, and sal. Nepal's and Bhutan's forests have silver fir and other conifers, oak, magnolia, beech, and birch hardwoods. Nepal's forests are threatened by overcutting and massive soil erosion. The government of Nepal is leading remedial conservation and reforestation projects.

DID YOU KNOW?

Sri Lanka has banned timber exports since 1977 to protect its rain forests.

Climate and Vegetation

The seasonal winds in South Asia heavily influence temperature and rainall. They also affect which crops people grow and how people and the environment are affected by not enough or too much rain.

Climate Regions

Most of South Asia is south of the Tropic of Cancer, so it has tropical climates with ample rainfall and diverse vegetation. Climates do vary in the north and west. In the north there are highlands in the Himalayas. In the west, there are deserts near the Indus River. When rain-bearing winds sweep in, hot climates burst forth.

Tropical Regions

India's west coast, Bangladesh, and southern Sri Lanka have tropical wet climates with a variety of vegetation. South Asia's rain forests absorb the moisture from seasonal rains blowing out of the southwest. Rain forests in western Sri Lanka, in southwestern India, and north of the Bay of Bengal have lush vines and orchids and ebony trees.

Surrounding the rain forest near the Western Ghats are coniferous and deciduous trees. Hot, damp Bangladesh has tropical forests with bamboo, mango, and palm trees.

The Sundarbans are threatened by rising sea levels and flooding saltwaters of the Bay of Bengal.

Midlatitude and Highland Regions

Cold Central Asian winds are blocked by the Himalayas. There is then a humid subtropical climate across Nepal, Bhutan, Bangladesh, and northeastern India. There are temperate mixed forests here.

South Asia is coldest in the Himalayan Mountains and Karakoram Mountains where there is snow year round. Little vegetation survives at high altitudes. In lower elevations, in the upper temperate region, are coniferous and hardwood trees. Bamboo stands and grasslands cover foothills.

Dry Regions

Northwestern India by the Indus River is arid, windswept desert land. East of the Indus in the Thar Desert thorny trees and grasses grow, and cattle graze. Indus irrigation aids in some wheat growth in the sandy soil.

Steppe surrounds desert, except on the coast. There are few trees in semiarid grassland. Between the Ghats in the Deccan Plateau of South India is steppe. Summer monsoons rise up in the Ghats, releasing rainfall moisture. Over the mountains, winds lose most of their moisture. This rain shadow effect makes the **leeward** mountainsides dry. India's interior is scrub and deciduous forest.

Seasonal Weather Patterns

Seasonal weather patterns bring badly needed rainfall to South Asia. Monsoon winds and other natural disasters bring catastrophe and hardship. The hot season with high temperatures is from late February to June. The cool season is October to late February. The mixed blessing of wet season with possible natural disasters is June to September.

Monsoon Rains

Monsoons are seasonal winds. In the hot season, hot air rises and becomes unstable. Moist Indian Ocean winds from the south and southwest change direction in the hot air bringing rainfall and flooding. In the cool season, it is reversed. Air from the interior of Asia cools, stabilizes, and blows south across the subcontinent to the ocean. The air is cool and dry.

East South Asia gets the heaviest monsoon rains. The Himalayas block rains from moving north, so they move west to the Gangetic Plains with rains for crops.

DID YOU KNOW?

India's 700 million farmers depend on monsoon rains and celebrate them when they come.

Natural Disasters

In India and Bangladesh, high temperatures and water permit farmers to grow needed rice crops. High temperatures without rain, such as those outside the monsoon paths in western Pakistan, bring drought and hunger while the Gangetic Plain receives downpours of rain and crops grow.

DID YOU KNOW?

Too much rain in low-lying Bangladesh causes flooding; kills people, livestock, and crops; and leaves people homeless.

Cyclones with high winds and heavy rains are a hazard in South Asia, especially Bangladesh. The 1999 cyclone in Orissa, India, had 160-mph winds and 20-foot waves that killed 10,000 people and caused more than $20 million in damages.

Tsunami

Indian Ocean and Himalayan Mountain tectonic activity affect India, Bangladesh, and Sri Lanka. The December 26, 2004 Indian Ocean tsunami was generated by the fourth most powerful earthquake recorded since 1900. Its epicenter was west of Sumatra. It measured 9.0 on the Richter scale. The 100-foot waves swept villages away, killed over 229,000 people in 12 countries, and destroyed farmlands, freshwater wells, coral reefs, mangrove forests, and coastal wetlands. The most damage was in South Asia and Southeast Asia.

In October 2005, an earthquake measuring 7.6 on the Richter scale killed 70,000 people and left millions without shelter and food in north Pakistan and Kashmir.

Practice

Insert T or F to indicate whether the statement is true or false, based on what you've read.

1. ______ India has mountains, deserts, and steppes.

2. ______ South Asian farmers celebrate the coming of monsoons.

3. ______ South Asia is safe from earthquakes and tsunamis.

4. ______ No vegetation is able to grow in India and Bangladesh.

India—Human Geography

The cultural history of India is thousands of years old. India is today a mix of ancient influences and modern cities, and is the world's largest democracy.

Population Patterns

The population of South Asia exceeds 1.1 billion people. That is more than 15% of the world population. Population density and distribution, in addition to urbanization, shape India's population patterns. The population is diverse.

People

The Dravidians lived in south India for 8,000 years. The Aryans came across Central Asia 3,000 years ago. The identity of many Indians is based on religion. They see themselves as Hindu, Muslim, Buddhist, Sikh, **Jain**, or Christian. Hindus define themselves by **jati**, or by occupation and social position.

Population Structure

India is growing so quickly it cannot provide basic food, education, and healthcare for its large, young, dependent population. Thirty-six percent of India's population is under 15 years old. In India, having few children is not a traditional value. Many people live on the streets; cities are overcrowded and too strained to provide adequate social services.

Economic development in India will mean higher wages and standards of living, education, and healthcare.

Density and Distribution

India's population density of 869 people per square mile is seven times the world average and varies by place. Climate, vegetation, and physical features affect how many people a place can support. Some monsoon-affected areas of India's southern coast and the fertile Gangetic Plain have 2,000 people per square mile, while the Thar Desert is sparsely populated.

Most of India's population (70%) is rural. They barely grow enough food for their families and some of their crops go to land owners. Cities have been growing as people migrate looking for jobs and better wages. Mumbai (Bombay) is India's main port on the Arabian Sea and its largest city, with 18.3 million people. Kolkata (Calcutta) is a port city on a branch of the Ganges specializing in iron and steel production. Delhi is part of a megalopolis closely linking metropolitan cities. Urban public resources and facilities are strained.

History and Government

India's ancient history still influences its population today. The Indus Valley civilization in Pakistan dates back 4,500 years. Today, India is integrating into the modern world.

First Civilizations

The Aryans, a group of hunters and herders from the northwest, settled in the Indus Valley around 2000 B.C. The Aryans created a rigid social structure

Europeans called the **caste system**. Social structure and religion are defined in the sacred Aryan writings called the *Vedas*. Religion defines India's history and culture.

Invasions and Empires

Other cultures followed the Aryans into northwestern India through the Khyber Pass in the Hindu Kush Mountains. The Mauryan Empire ruled India from 320 to 180 B.C. The Gupta Empire governed central India from A.D. 320 to 500 and became one of the most advanced world civilizations. The Muslim Moghul Empire converted many Indians to Islam.

Most of South Asia was ruled by the British Empire. As mercantilists, they used colonies to supply materials and markets to the empire. The British practiced political and economic **imperialism**, or domination. The British **raj** (the Hindu word for *empire*), introduced English, restructured the educational system, built railroads, and developed a civil service.

Independence

India's fight for independence was led by Mohandas K. Gandhi. Gandhi advocated nonviolence in his public statements and actions for Indian self-rule. In 1947, the British colony of India was divided into mostly Muslim Pakistan and mostly Hindu India.

Education and Healthcare

Children in India attend school to age 14. The literacy rate is 61%, but lower in rural areas. The government strives to provide educational opportunity to women and the poor. Urban schools require uniforms and attendance two Saturdays a month.

India's state-run hospitals are improving. Malaria and other diseases are now controlled. HIV and AIDS are still a problem.

Language and Religion

The people of India speak 18 official languages and hundreds of dialects. Hindi is most widely spoken and English, the language of international business and tourism, is spoken in former British areas. Government documents are written in Hindi and English.

Hindu is the main, **polytheistic** religion of India. Islam, Christianity, Buddhism, and Sikhism are practiced. Sikhism, founded by the **guru**, or teacher, Nā nak, is monotheistic and stresses good deeds and meditation. Most Sikhs live in northwest India and want an independent state.

Arts

The *Mahābhārata* and *Rāmāyana*, two great epic poems, mix Hindu and social beliefs. India has many classical dance styles mostly based on themes from Hindu mythology. Indians have loved movies since the first ones arrived in 1896. Bollywood (a portmanteau of Bombay and Hollywood) produced 1,041 full-length films in 2005. Hollywood made 699 full-length films in the same year. Bollywood is the largest film center in the world.

Family Life and Leisure

Family is the most important social unit in India with **extended families** living together, sharing household chores and finances. A bride will go to live with her husband and his extended family. The traditional arranged marriage based on caste, economic status, and education is slowly changing. Within the family, influence is based on gender and age, and for women, the number of male children they produce. Movies and television are important leisure activities for the middle class and rural people attend weddings and family celebrations.

Pakistan and Bangladesh

Histories of Pakistan and Bangladesh have included horrible conflict, but also great cultural achievement. Both have a history of Muslim influence, British colonial rule, and the pursuit of independence. The war on terror has made the world more familiar with Pakistan's legacy of hardship.

Population Patterns

Population growth and movement have deeply impacted Pakistan and Bangladesh. Most people in Pakistan and Bangladesh share Islam and a rural existence.

People

Punjabis, Sindhis, Pashtuns, Mohajirs, and Baluchis are Pakistan's major ethnic groups. Ethnic identity is based on ethnic background, language, and religion. Most people in Bangladesh are Bengali, like most people in the Hindu Indian state of Bengal. Most people in Bangladesh, however, are Muslim.

Density and Distribution

South Asia's most densely populated area is Bangladesh, with 2,596 people per square mile. Dhaka is now the third most densely populated city in the world. Although Bangladesh has rich soil and improved farming techniques, it has a hard time feeding its population. Fertility rates have decreased since women's education has improved. The Bengali government and private groups have provided women with small business loans.

Pakistan is South Asia's most urbanized country, with 34% of its population living in the city. Urban populations in Pakistan are growing. There are job and housing shortages and pollution in cities. Rural Pakistanis are attracted to Pakistan's modern capital Islamabad and its booming port city Karachi.

History and Government

Similarities and differences have impacted the histories of Pakistan and Bangladesh.

Indus Valley Civilization

About 2500 B.C., the Indus River valley civilization developed a writing system, a strong central government, a thriving world trade, and some of the world's first cities with indoor plumbing at Mohenjo Daro and Harappa. Environmental changes like flooding or drought destroyed these cities around 1750 B.C.; then the Aryans moved into the area.

Islam's Impact

Muslim invaders and traders settled in southeast Pakistan around the A.D. 700s. Muslim teachers converted many people to Islam in what is now Bangladesh. Islam ruled both areas from the 1500s to the 1800s. Parts of Pakistan and India were conquered in the nineteenth century by Sikhs, who combined Islam and Hinduism. The British took back the land.

The idea of a separate Muslim state emerged in the 1930s. Since the Muslims and Hindus could not agree on a constitution, the British granted independence to India and Pakistan. Pakistan consisted of East and West Pakistan, separated by 1,000 miles of Indian land.

Conflicts and Government

North of India and Pakistan, Kashmir was ruled for centuries by Indian princes, or **maharajas**. When Kashmir was declared part of Pakistan, the Hindu prince fled to Delhi and signed his state over to India. India claimed legal right to Kashmir, but Pakistan insisted Pakistan was a better ruler for the majority of its Muslim residents.

There have been wars and fighting between India and Pakistan for decades. In 1998, India, then Pakistan, conducted underground nuclear weapons tests.

People in East Pakistan are Muslim, but they are ethnic Bengali and speak the language Bangla. Following independence, West Pakistan wanted to impose its language, Urdu, on all of Pakistan. Bengalis in the east protested.

In 1970–1971, Bengali nationalists declared independence from Pakistan after a difficult civil war, naming the country Bangladesh, which means "Bengal country." Bangladesh is a parliamentary republic. Violence and political and ethnic rivalry make stable rule difficult.

Pakistan is a parliamentary republic plagued with instability and military rule. In 1999, charges of corruption led a military coup against Prime Minister Navez Sharif, led by General Pervez Musharaf. In 2008, Musharraf resigned and is currently in exile.

Education and Healthcare

Education in Pakistan and Bangladesh lags behind the rest of South Asia. The literacy rate is 49% in Pakistan and 48% in Bangladesh. Most children attend elementary school in Bangladesh, but only 47% go on to secondary school. Only 36% of females are literate in Pakistan. In some parts of Pakistan, educating girls is prohibited on religious grounds.

Both countries have very poor healthcare. In 2005, an earthquake near Islamabad destroyed 85% of the city's infrastructure, including healthcare facilities and schools. The people of Bangladesh face the serious threat of water-borne disease since the country lies in a vast flood plain.

Language and Religion

The main language of Bangladesh is Bangla. Even though the official language of Pakistan is Urdu, only 8% of the people there speak it; Punjabi is the dominant language of Pakistan. People from Pakistan and Bangladesh who attend university or work in government speak English.

Islam is the main religion in Pakistan and Bangladesh. Hinduism, Buddhism, and Christianity are practiced in both countries. Sikhism is practiced in northern Pakistan.

Arts

South Asians have used visual arts to express religious beliefs and document daily life. Stone carving and sculpture date back to the Indus Valley civilization. Mauryan Empire techniques for marble polishing have never been duplicated. Mogul emperors relaxed Muslim prohibitions against representing human forms and they commissioned portraits and decorative paintings.

Literature and dance are important in Bangladeshi culture. In 1913, Bengali Rabindranath Tagor became the first non-European writer to win the Nobel Prize for Literature. The plays and poetry of Kazi Nazrul Islam, "the voice of Bengali nationalism," have inspired farmers for decades with their political and historical themes about the oppression of Muslims. Bangladesh is also home to creative indigenous dances.

The richest Pakistani art forms are music and literature. A form of devotional singing called *Qawwali* is very popular. At public *musha'irahs* people recite poetry. The traditional rhythmic form of *raga* can be traced back to the thirteenth-century poet and musician Amir Khosrow.

Family Life

The center of social life in Pakistan and Bangladesh is family. Extended families are nearby and multiple households often share a home. Many marriages are arranged, but educated men and women are starting to choose their own partners. A new wife usually lives with the husband's family. Urban areas have smaller families.

Nepal, Bhutan, Maldives, and Sri Lanka

Population Patterns

The mountainous northeast and island cultures of South Asia demonstrate centuries of influence from surrounding regions. Ethnic diversity creates fascinating cultures, but they can clash.

People

Nepal was influenced by migrations from the north in Tibet. The Sherpas are master Tibeto-Nepalese mountaineers. Nepal also had migrations from the Indus region to the south.

The majority of people in northern, central, and western Bhutan are descendants of Tibetans called the Bhote. The Bhote and the Sharchrops in eastern Bhutan speak Tibetan dialects and practice Tibetan Buddhism. In southern Bhutan, 35% of the population is Nepalese. Called the Gurung, they speak Nepalese and practice Buddhism.

The islands of Maldives have people from southern India, Sri Lanka, East Africa, and Arab countries. Southern Indians probably settled there first, then Sri Lankan, East African, and Arab sailors.

On the island of Sri Lanka, the Buddhist Sinhalese are in the majority and control the government. The Hindu Tamils have been fighting for an independent Tamil state in the north since the early 1980s. Since 1984, more than 60,000 Sri Lankans have been killed or disappeared.

Density and Distribution

In Nepal and southern Bhutan, population density is between 55 and 447 people per square mile. In the north, population decreases with elevation. There are only 25 people per square mile in the Himalayas due to the harsh climate. The most densely populated area of Nepal is the Kathmandu Valley.

There are 778 people per square mile in Sri Lanka. The 1,190 tiny coral islands of the Maldives are crammed full of 2,586 people per square mile.

History and Government

India, Arab countries, and European powers, as well as internal influences, have shaped the histories of Nepal, Bhutan, Sri Lanka, and Maldives.

Early History

In A.D. 400, the Licchavi dynasty was established in Nepal's Kathmandu Valley. They likely came from India. Around the ninth century, three new dynasties replaced the Licchavi dynasty, which ceased to exist. Under the new Shah dynasty, Nepal expanded southward in the 1700s and confronted the British in a two-year Anglo-Nepalese war (1814–1816). Nepal was reduced to its current size, but remained independent.

Not much is known of Bhutan prior to Tibetan Buddhist monks arriving there in the A.D. 800s. Bhutan's political and religious history is linked to the monasteries and monastic schools.

The Sinhalese arrived in Sri Lanka from India during the 500s B.C. Arab, Greek, and Roman sailors glimpsed the Sinhalese civilization from 200 B.C. to A.D. 1200.

Buddhists from South Asia first settled the Maldive Islands. In the twelfth century, Islam arrived on the Maldives. Portuguese traders came to the Maldives in the fourteenth century.

Path to the Modern

In the early 1600s, a Tibetan **lama**, or Buddhist monk, consolidated political and religious power in Bhutan by developing a system of law. After his death, civil war and chaos plagued Bhutan until the 1800s, when a ruler established links to the British in India. India and Britain respected Bhutan's sovereignty. Bhutan is now evolving into a constitutional monarchy with representative government.

In the late 1700s, a ruler of a small principality combined states to form Nepal. Nepal experienced periods of internal political strife before achieving stability. Because of its isolation, Nepal was not colonized by European powers. Nepal has seesawed between representative government and monarchical rule with a king as head of state. The country's ruling party and rebel groups continue to discuss peace.

Since the 1500s, Portuguese, Dutch, and British powers fought to control Sri Lanka's very profitable position on the spice trade route. Portugal and Spain entered the South Asian maritime trade in the 1500s. The Dutch became dominant in the 1600s. The British were dominant in the 1700s. Although the British came to rule Sri Lanka, Dutch law remains. The British developed plantation economies for tea, rubber, and coconuts.

Sri Lanka is a parliamentary republic. Over 30,000 Sri Lankans were killed in the 2004 underwater earthquake and tsunami. Millions of people lost everything and needed water, food, shelter, and medical care. Disputes between the Sri Lankan government and the now-defunct Hindu Tamil Tigers in the north delayed delivery and distribution of international aid to the needy.

Since the 1100s, Arab and Muslim influence has been strong in the Maldive Islands. Sultans have ruled the islands most of the time. In the 1500s, the Portuguese ruled the Maldives briefly. From the late 1800s to 1965, the British ruled the Maldives. In 1965, the Maldives became an independent republic.

Language and Religion

In Nepal, most people are Hindu. A small number of people in Sri Lanka also practice Hinduism. In Sri Lanka and Bhutan, Buddhism dominates. It is also practiced in Nepal. In Bhutan and Nepal, people fly colorful prayer flags and prayer wheels sending out invocations twirl in many places. Monks chant or repeat prayers called **mantras**. The Maldives was founded as a Muslim state when traders began commerce on the islands.

The languages of Nepal, Sri Lanka, and Maldives have Indo-European roots. In Nepal most people speak Nepali, but twelve other languages are spoken there. Sri Lankans speak Sinhalese, the official language, and Tamil, whose roots are Dravidian from southern India. People of the Maldives speak Dhivehi and English. People from Bhutan speak Dzongka of Tibetan origin.

Education and Healthcare

Education levels in South Asia are uneven. In the Maldives, where literacy reaches 93%, traditional schools teach the Quran and English-language schools teach children through the secondary level. Sri Lankans receive free education from primary school through university. Only 20% of elementary-age children in Bhutan are enrolled in school. Bhutan's literacy rate is 47% and Nepal's is 48%.

The availability and quality of healthcare is low. In the Maldives, people rely on traditional medicine. Hospitals and healthcare professionals are few in these countries, especially in rural areas. Short life expectancies are often the result, although Sri Lanka does have a 73-year life expectancy, close to that of the United States. With scarce clean water, water-borne diseases like cholera and dysentery are common.

Arts

The spirit of art is seen every day all across South Asia. Buddhist **stupas**, or domed shrines, of Nepal and Sri Lanka and the fortified monasteries, or **dzong**, of Bhutan reveal this spirit. Dance is popular in Nepal and Bhutan and it often incorporates religious and historical stories. In literature, Michael Ondaatje, born in Sri Lanka and now living in England, won England's prestigious Booker Prize for his novel *The English Patient*, which became an Academy Award-winning film.

Economy

Economic growth has occurred in different ways and at different rates in the countries of South Asia. Increased trade has made South Asia's countries more economically interdependent.

Agriculture is the primary economic activity in South Asia. A variety of economic activities contribute to South Asia's growth.

Agriculture

About 60% of the labor force in South Asia is employed in agriculture. Many people practice traditional subsistence farming and grow just enough food to survive. Many people use hand plows and digging sticks and sow seeds by hand. Some people have irrigation.

They use oxen and water buffalo to pull wood plows, turn water wheels, or power mills for grinding grain. They may use long-haired yaks accustomed to high elevations, camels in deserts, and elephants instead of tractors.

Agricultural Conditions

Farms in South Asia vary in size depending on geographic, historic, and cultural factors. In the highlands, terracing is used on the sleep slopes to carry water downhill. There are fruit orchards in the fertile highlands of Pakistan. In Bangladesh's deltas and along South Asia's rivers, farmers wade knee-deep in water to grow rice. Farms in India are small; British, Dutch, and Dutch colonizers established huge, technically advanced tea, rubber, and coconut plantations in Sri Lanka.

Crops

Plantations in India grow cash crops like bananas, tobacco, coffee, and tea for domestic sale or export. Cotton is the big cash crop in India and Pakistan. Bangladesh's cash crop is **jute** fiber used to make rope, string, and cloth. Rice grows in the tropical wet climate of the Ganges; India is second only to China in rice production.

Agricultural Improvements

Since the 1960s, the **Green Revolution** has sought ways to increase crop yields in developing countries using managed irrigation, fertilizers, and yield varieties of crops. Modern irrigation and mechanization is costly, so when short on petroleum, plant material or biomass fuel (animal dung) is used for energy.

> **DID YOU KNOW**
>
> Supporters of the Green Revolution say it relieves hunger. Opponents say chemicals pollute water and build up insect resistance to chemical fertilizers.

Mining and Fishing

Mining and fishing continue to be profitable in South Asia. The Gangetic Plain and the mountains of east India are rich in minerals like iron, coal, copper, and bauxite. Bhutan has lead and marble. Sri Lanka has graphite and precious and semiprecious stones. Several South Asian countries have petroleum and natural gas.

Bordered in part by oceans and watered by rivers, South Asia has rich fishing resources that bring needed income. Pakistan and Sri Lanka export shrimp and lobster.

> **DID YOU KNOW?**
>
> Two thirds of Sri Lanka's fishing fleet was destroyed in the 2004 tsunami.

Transportation and Communications

Transportation and communications vary across South Asia. Nepal has footpaths and India a vast rail system.

Land Travel

Nepal has the least developed transportation system in South Asia. India has a 150-year-old rail system developed by the British and 20,000 miles of highways to all major cities. Bangladesh has a British rail system and roads with cycle risckshaws and buffalo carts. Pakistan and Bhutan have recently invested in highway construction.

Water Travel

South Asian countries, aside from landlocked Bhutan and Nepal, have many seaports linking major ocean trade routes. Mumbai, Kolkata, and Chennai handle India's trade. Karachi and Port Qasim handle

Pakistan's trade. Inland waterways are important to Bangladesh.

People and Their Environment

South Asian countries are looking for ways to manage natural resources. These resources are in danger of depletion due to exploding population growth, high population densities, and a growing middle class.

Countries in South Asia are attempting to manage their natural resources in ways that do not deplete them or cause more damage to ecosystems. **Sustainable development**, using resources at a rate that does not deplete them, is what South Asia needs.

Water

Lack of clean water is a persistent problem in South Asia. Even in India, the most developed country in the region, only 54% of urban dwellers and 21% of rural folk have access to sanitation facilities. Most people in the region must use water polluted by human waste and chemical runoff.

Due to climate variations in South Asia, northwest India may face drought while Bangladesh deals with flooding. Dams can control flooding and hold reserve water for drought. Unfortunately, though, dams trap soil nutrients, flood some areas and displace villagers, and disrupt the balance of the ecosystem. Reservoirs breed malaria-carrying mosquitoes.

The building of dams in central India's Narmada River basin has been plagued by 25 years of controversy—millions of acres subject to drought can be irrigated and hydroelectric power created, but ancestral villages would be flooded and people would have to be resettled.

Forests

India was covered by forests. Commercial timber and mining operations, human settlement, slash-and-burn agriculture, and burning biomass for fuel destroyed old-growth forests. Destroying the mangrove forests of Bangladesh's Sundarbans by the Ganges River delta weakens protection from cyclones and erosion. The destruction of mangrove forests in Sri Lanka made the 2004 tsunami much more devastating.

Tropical rain forest tree roots trap precipitation, keep soil in place, and cool temperatures. Destroying these forests erodes soil, causes rain flooding, and raises temperatures.

Gandhi's follower Sunderlal Bahaguna built on the traditional South Asian respect for trees with his **chipko** "tree hugger" movement. Bahaguna has reminded villagers of the importance of trees and provided nurseries with seedlings for reforestation.

Wildlife

Deforestation and irrigation destroy the amazing variety of wildlife habitats in South Asia. They also drive wildlife into contact with humans. India's and Sri Lanka's rain forests have elephants, monkeys, and water buffalo. Bengal tigers and crocodiles roam Bangladesh. Colonials and farmers have killed animals to protect crops and livestock.

South Asian governments and conservation groups have created wildlife reserves and written laws against hunting and logging. Poaching of protected animals is challenged with economic reasons to protect wildlife.

Solution Seeking

Industrialization causes air pollution. The four most polluted cities in the world, Mumbai, New Delhi, Chennai, and Kolkata, are in India. To protect against monsoon destruction, meteorologists studying weather patterns research how to predict how intense approaching storms in the Bay of Bengal are and when they might hit.

DID YOU KNOW?

Satellite imaging is helping scientists understand erosion caused by flooding in Bangladesh's coastal deltas that displaces families each year.

Future Challenges

Continuing conflicts impact the people and environments of South Asia. Ethnic and religious diversity in South Asia often result in conflict. Minority Hindus or Muslims on the wrong side of the India-Pakistan border still face violence.

Conflict in Kashmir

India and Pakistan still fight over who should control mostly Muslim Kashmir. They have had three wars over the territory. Pakistan today controls Kashmir's northwestern portion while the rest is held by India. Troops from both countries patrol the border India wants internationally recognized. Economic ties and trade agreements created in 2005 may steer India and Pakistan away from the kind of armed conflict that occurred in 2002.

India and Pakistan tested nuclear warheads in 1998. The escalation of nuclear capabilities, or **nuclear proliferation**, costs both countries money people need for food, housing, education, and health care. International sanctions leveled against India and Pakistan have made worse the hardships people of the region face.

Practice

Fill in the blank with the correct word from the information in the preceding paragraphs.

5. India's majority religion is ____________.

6. The religion of most people in Pakistan and Bangladesh is ____________.

7. Rain forest tree roots stop ____________.

8. Government documents in India are written in ____________ and English.

9. __________ is South Asia's most urbanized country

Answers

1. T
2. T
3. F
4. F
5. Hinduism
6. Islam
7. Erosion
8. Hindi
9. Pakistan

POSTTEST

Now that you have completed all the lessons, it's time to show off your new skills. Take the posttest in this chapter to see how your geography knowledge has improved. This test has 50 multiple-choice questions covering the topics you studied in this book. While the format of the posttest is similar to that of the pretest, the questions are different.

After you complete the posttest, check your answers with the key at the end of the chapter. If you still have weak areas, go back and work through the applicable lessons again.

1.	a	b	c	d
2.	a	b	c	d
3.	a	b	c	d
4.	a	b	c	d
5.	a	b	c	d
6.	a	b	c	d
7.	a	b	c	d
8.	a	b	c	d
9.	a	b	c	d
10.	a	b	c	d
11.	a	b	c	d
12.	a	b	c	d
13.	a	b	c	d
14.	a	b	c	d
15.	a	b	c	d
16.	a	b	c	d
17.	a	b	c	d
18.	a	b	c	d
19.	a	b	c	d
20.	a	b	c	d
21.	a	b	c	d
22.	a	b	c	d
23.	a	b	c	d
24.	a	b	c	d
25.	a	b	c	d
26.	a	b	c	d
27.	a	b	c	d
28.	a	b	c	d
29.	a	b	c	d
30.	a	b	c	d
31.	a	b	c	d
32.	a	b	c	d
33.	a	b	c	d
34.	a	b	c	d
35.	a	b	c	d
36.	a	b	c	d
37.	a	b	c	d
38.	a	b	c	d
39.	a	b	c	d
40.	a	b	c	d
41.	a	b	c	d
42.	a	b	c	d
43.	a	b	c	d
44.	a	b	c	d
45.	a	b	c	d
46.	a	b	c	d
47.	a	b	c	d
48.	a	b	c	d
49.	a	b	c	d
50.	a	b	c	d

Directions: Select the letter of the best answer and circle it.

1. The Earth's atmosphere is made of mostly ___________.
a. oxygen
b. carbon dioxide
c. nitrogen
d. helium

2. The Tropic of ___________ at 23.5° N is the northernmost point to receive direct sunlight.
a. Cancer
b. Capricorn
c. Equinox
d. Solstice

3. A(n) ___________ is a community of plants and animals dependent on one another and their surroundings to survive.
a. biosphere
b. resource
c. ecosystem
d. physical geographic process

4. ___________ geography focuses on history, government, population growth, urban development, economic production and consumption, the arts, healthcare, and education.
a. Physical
b. Habitat
c. Human or cultural
d. Transitional

5. Birth rates are high in the ___________ world.
a. developed
b. industrial
c. Western
d. developing

6. Seas, gulfs, and bays are small ___________ water bodies.
a. fresh
b. salt
c. lake
d. tributary

7. Nearly all Earth's surface water is ___________ water.
a. ground
b. fresh
c. salt
d. river

8. The lithosphere is Earth's ___________.
a. landforms
b. mantle
c. crust
d. core

9. The ___________ of Fire is an area of earthquake and volcanic activity that crosses continents and oceans.
a. Magma
b. Trench
c. Slope
d. Ring

10. As glaciers melt and recede, they leave large piles of rock and debris known as ___________.
a. icebergs
b. glacial lakes
c. moraines
d. wind erosions

11. The Grand ___________ was formed by water erosion.
a. Tsunami
b. Tetons
c. Colombia
d. Canyon

12. Lines of ___________, or parallels, circle Earth in degrees horizontally along the equator.
a. latitude
b. longitude
c. legend
d. meridians

13. Lines of longitude, or ___________, circle Earth vertically from Pole to Pole.
a. meridians
b. prime meridians
c. international date lines
d. time zones

14. Southwest Asia, North Africa, and Central Asia are rich in ___________.
a. timber
b. water
c. produce
d. petroleum

15. Indigenous North Africans preceding Arab invasions were ___________.
a. Maoris
b. Berbers
c. Bedouins
d. Ottoman Turks

16. Judaism, Christianity, and Islam were born in the ___________.
a. Eastern Mediterranean
b. Arabian Peninsula
c. Sahel of North Africa
d. Central Asian steppe

17. ___________ largest agricultural floodplain surrounds the Ganges River.
a. India's
b. China's
c. Pakistan's
d. Nepal's

18. The cities of Quebec, Montreal, and Ottawa developed along the ___________.
a. Mississippi River
b. Great Plains
c. St. Lawrence River
d. Gulf of Mexico

19. Since the 1970s, the mild climates of the American ___________ and Southwest have attracted rapid population growth.
a. Midwest
b. Northeast
c. Rust Belt
d. South

20. The ___________ climate zone of South America is coldest.
a. tierra helada
b. puna
c. tierra fria
d. tierra templada

21. The Olmec, Maya, Inca, Maori, Australian, Native American, and other indigenous tribes are also known as ___________ cultures.
a. aboriginal
b. mestizo
c. machismo
d. Creole

22. The largest country in the world is ___________.
a. Russia
b. China
c. India
d. Indonesia

23. Ample coniferous forests make up Russia's largest climate zone, the ___________.
a. tundra
b. subarctic
c. taiga
d. humid continental

24. Not including migration, negative population growth simply means ___________.
a. the birth rate exceeds the death rate
b. the death rate exceeds the birth rate
c. the birth rate is not recorded
d. birth rate and death rates are equal

25. Glaciation formed the fjords of ___________.
a. the Mediterranean
b. Scandanavia
c. canals
d. the North Sea

26. England, Wales, Scotland, and Northern Ireland make up ___________.
a. the United Kingdom
b. the Celts
c. the British Isles
d. Great Britain

27. The ___________ now combine(s) most of Europe into one economic community and conducts more trade by volume than any country in the world.
a. North Atlantic Treaty Organization (NATO)
b. Organization of Petroleum Exporting Countries (OPEC)
c. European Union (EU)
d. guest workers

28. One positive response to deforestation is ___________.
a. replanting trees and forests
b. expanding industrial use of coal
c. a and **b**
d. none of the above

29. The edges of African plateaus are marked by steep, jagged cliffs called ___________.
a. cataracts
b. escarpments
c. estuaries
d. kums

30. South Africa has about half the world's ___________.
a. rice
b. gold, diamonds, and uranium
c. spices
d. bat guano

31. Large parts of Africa south of the Sahara currently face ___________.
a. drought, famine, lack of clean water, inadequate sanitation, and epidemic disease
b. corruption, civil war, and human rights violations
c. a and **b**
d. none of the above

32. Independence from European rule came most recently in ___________.
a. Latin America
b. Asia
c. Africa
d. the Middle East

33. Racial segregation in South Africa was called ___________.
a. apartheid
b. Afrikaner
c. Zimbabwe
d. Angola

34. Mount Fuji is in ___________.
a. the Himalayas
b. China
c. Japan
d. Taiwan

35. The Himalayas separate China from ___________.
a. Japan
b. Taiwan
c. South Asia
d. Mount Everest

36. Seasonal monsoon winds help the major crop of ___________ to grow in Southeast Asia's fertile paddies.
- **a.** rice
- **b.** wheat
- **c.** corn
- **d.** livestock

37. Where is population density lowest?
- **a.** China
- **b.** Japan
- **c.** Singapore
- **d.** Mongolia

38. ___________ is the world's most populous urban area, with more than 35 million people.
- **a.** Beijing
- **b.** Tokyo
- **c.** Rio de Janiero
- **d.** Shanghai

39. Since 1945, Communist ___________ Korea has sought to unify Korea and expand its nuclear capabilities.
- **a.** North
- **b.** South
- **c.** West
- **d.** East

40. Over 60% of high-tech imports to the United States come from ___________ Asia.
- **a.** South
- **b.** Southeast
- **c.** East
- **d.** Central

41. The ___________ Peninsula includes all of Vietnam, Laos, Cambodia, Myanmar, and part of Thailand.
- **a.** Korean
- **b.** Indochina
- **c.** Hong Kong
- **d.** Malay

42. ___________ cultivates over 1,000 species of orchids.
- **a.** Thailand
- **b.** Vietnam
- **c.** Indonesia
- **d.** Laos

43. In the highland climate of Myanmar, you can find ___________.
- **a.** mangrove trees
- **b.** cypress trees
- **c.** rhodendrons
- **d.** moss

44. ___________ cities like Bangkok serve as a country's port, economic center, and often the capital.
- **a.** Primate
- **b.** Viceroyalty
- **c.** Megalopolis
- **d.** Animism

45. The temple complex of Angkor Wat in Cambodia was influenced by ___________.
- **a.** Hinduism
- **b.** Koryoism
- **c.** Islam
- **d.** Roman Catholicism

46. Indonesia, Thailand, and Malaysia lead the world in ___________ production.
- **a.** tea
- **b.** bamboo
- **c.** rubber
- **d.** petroleum

47. Trawlers are used in commercial ___________.
- **a.** cattle ranching
- **b.** rice farming
- **c.** timber cutting
- **d.** fishing

48. The June 1991 volcanic eruption of Mount Pinatubo in ___________ was one of the twentieth century's most violent and destructive volcanic eruptions.
a. Indonesia's island of Bali
b. Indonesia's island of Java
c. the Philippines
d. Japan

49. The world's largest coral reef lies off the coast of ___________.
a. New Zealand
b. South Africa
c. Australia
d. the United States

50. Most animal life in Antarctica, such as whales, seals, and penguins, can be found ___________.
a. in the Antarctic Ocean
b. in the Vinson Massif
c. in East Antarctica
d. at the South Pole

ANSWERS

1. c.
2. a.
3. c.
4. c.
5. d.
6. b.
7. c.
8. a.
9. d.
10. c.
11. d.
12. a.
13. a.
14. d.
15. b.
16. a.
17. a.
18. c.
19. d.
20. a.
21. a.
22. a.
23. b.
24. b.
25. b.
26. a.
27. c.
28. a.
29. b.
30. b.
31. c.
32. c.
33. a.
34. c.
35. c.
36. a.

37. d.
38. b.
39. a.
40. c.
41. b.
42. a.
43. c.
44. a.
45. a.
46. c.
47. d.
48. c.
49. c.
50. a.

GLOSSARY

Aboriginal being the first or earliest known inhabitants of a region

absolute location the exact position of a place on Earth's surface

accretion the slow process of a sea plate sliding under a continental plate, creating debris that can cause continents to grow outward

acculturation the cultural modification of an individual, group, or people by adapting to or borrowing traits from another culture

acid depositions wet or dry airborne acids falling to Earth

acid rain precipitation carrying large amounts of dissolved acids damaging buildings, forests, and crops, and killing wildlife

alluvial plain a floodplain, such as the Gangetic Plain in South Asia, on which flooding rivers have deposited rich soil

alluvial soil rich soil made of sand and mud deposited by running water

altiplano Spanish for "high plain"; a region in Peru and Bolivia encircled by the Andes Mountains

animism the belief that spirits inhabit natural objects and forces of nature

apartheid the policy of strict separation of the races adopted in South Africa in the 1940s

aquaculture the cultivation of fish and other seafood

aquifer underground water-bearing layers of porous rock, sand, or gravel

arable suitable for growing crops

archipelago a group or chain of islands

artesian well a bored well from which water flows up like a fountain

Asia-Pacific Economic Cooperation (APEC) a trade group whose members ensure that trade among the member countries of Asia and the Pacific is efficient and fair

Association of Southeast Asian Nations (ASEAN) an organization formed in 1967 to promote regional development and trade in Southeast Asia

atheism the belief there is no God

atmosphere a layer of gases surrounding Earth

atoll a ring-shaped island formed by coral building up along the rim of an underwater volcano, usually with a central lagoon

autocracy a government in which one person rules with unlimited power and authority

autonomous areas minor political subunits created in the former Soviet Union and designed to recognize the special status of minority groups within existing republics

avalanche a large mass of ice, snow, or rock that slides down a mountainside

axis an imaginary line that runs through the center of the earth between the North and South Poles

balkanize to divide a region into smaller regions, often hostile toward each other

Bedouin a member of the nomadic desert peoples of North Africa and Southwest Asia

bilingual speaking or using two languages

biofuels energy sources derived from plants or animals. Throughout the developing world, wood, charcoal, and dung are primary energy sources used for cooking and heating

biomass plant and animal waste used especially as a source of fuel

biome ecologically interactive flora and fauna adapted to a specific environment. Examples: a desert or a tropical rain forest

biosphere the part of Earth where life exists

birth rate the number of births per year per 1,000 people

black market the illegal trade of scarce or illegal goods, usually sold at high prices

blizzard a snowstorm with winds of more than 35 miles per hour, temperatures below freezing, and visibility of less than 1,320 feet for 3 hours or more

Bolshevik a member of the communist party that seized power in Russia in the Revolution of November 1917

boomerang a curved throwing stick used by Aborigines for hunting in Australia

boreal forest coniferous forest found in high-latitude or mountainous environments of the Northern Hemisphere.

brain drain the loss of highly educated and skilled workers to other countries

buffer state neutral territory between rival powers

campesinos farm workers or people who live and work in rural areas

canopy the top layer of a rain forest, where the tops of tall trees form a continuous layer of leaves

carrying capacity the population that an area will support without undergoing deterioration

cartogram a map that presents statistical data by geographic distribution

cartography the science of mapmaking

cash crop farm products grown to be sold or traded rather than used by the farm family

caste system the complex division of South Asian society into different hierarchically ranked hereditary groups. Most explicit in Hindu society, but also found in other cultures to a lesser degree

cataract a large waterfall

caudillo a Latin American political leader or strongman from the late 1800s to the present, who is often a military dictator

central business district the traditional business and commercial center of a city or town, sometimes referred to as downtown

cereal any grain such as barley, oats, or wheat, grown for food

channel a long gutter, groove, or furrow

chaparral a type of vegetation made up of dense forests of shrubs and short trees, common in Mediterranean climates

chernozem rich, black topsoil found in the Northern European Plain, especially in Russia and Ukraine

chinampas floating farming islands made by the Aztec

chinook seasonal warm wind blowing down the Rockies in late winter and early spring

chipko India's "tree hugger" movement protecting forests through reforestation and by supporting limited timber production

chlorofluorocarbon chemical substance, found mainly in liquid coolants, that damages Earth's protective ozone layer

city-state an independent community consisting of a city and the surrounding lands

civil of or relating to citizens

clan a tribal community or large group of people related to one another

clear-cutting the removal of all trees in a stand of timber

climate weather patterns typical for an area over a long period of time

Cold War the power struggle between the Soviet Union and the United States after World War II

collectivization the combining of small, privately owned agricultural parcels into larger, state-owned farms. This was a central component of communism in Eastern Europe and the Soviet Union

colonialism the formal, established rule over local peoples by a larger, imperialist government; the expansion of political and economic empire

Columbian exchange an exchange of people, diseases, plants, and animals between the Americas and Europe/Africa initiated by the arrival of Christopher Columbus in 1492

command economy a system of resource management in which decisions about production and distribution of goods and services are made by a central authority

commercial farming agriculture or aquaculture organized as a business

commodities goods produced for sale

Commonwealth of Independent States (CIS) a loose political union of former Soviet republics established in 1992 after the dissolution of the Soviet Union

commune a collective farming community whose members share work and products

communism a totalitarian system of government

compass rose a map tool that indicates direction

compound something formed by a union of elements or parts

condensation the process of excess water vapor changing into liquid water when warm air cools

coniferous vegetation having cones and needle-shaped leaves, including many evergreens that keep their foliage throughout winter

conquistador the Spanish term for "conqueror," referring to soldiers who conquered Native Americans in Latin America

conservation farming a land management technique that helps protect farmland

consumer goods products and services that directly satisfy human wants

continental drift the theory that the continents were once joined and then slowly drifted apart

continentality the effect of extreme variations in temperature and very little precipitation within the interior portions of a landmass

continental shelf the part of a continent that extends underwater

cooperative a voluntary organization whose members work together and share expenses and profits

coral limestone deposits formed from the skeletons of tiny sea creatures

cordillera parallel chains or ranges of mountains

core innermost layer of Earth. The core is made up of a super-hot but solid inner core and a liquid outer core

Coriolis effect the sideways movement of prevailing winds caused by Earth's rotational movement and speed

cottage industry businesses that employ workers in their homes

coup d'état a violent overthrow of the government

crude oil unrefined petroleum

Crusades a series of religious wars (AD 1100–1300) in which European Christians tried to retake Palestine from Muslim rule

crust the rocky shell forming Earth's surface

cultural convergence the mixing of cultures

cultural diffusion the spread of new knowledge and skills from one culture to another

cultural divergence the separation of people or societies, with regard to beliefs, values, and customs, because of distinctly different political systems

cultural hearth a center where cultures developed and from which ideas and traditions spread outward

culture the way of life of a group of people who share beliefs and similar customs

culture region a division based on a variety of common factors, including government, social groups, economic systems, language, or religion

cuneiform Sumerian writing system using wedge-shaped symbols pressed into clay tablets

current a cold or warm stream of seawater that flows in the oceans, generally in a circular pattern; an interval of time during which a sequence of a recurring succession of events or phenomena is completed

cyclone a storm with heavy rains and high winds that blow in a circular pattern around an area of low atmospheric pressure

czar the ruler of Russia until the 1917 revolution

dalits the "oppressed." In India, people assigned to the lowest social class

death rate the number of deaths per year for every 1,000 people

deciduous falling off or shed seasonally or periodically; trees such as oak and maple, which lose their leaves in autumn

decolonialization the process of a former colony gaining (or regaining) independence over their territory and establishing (or reestablishing) an independent government

deforestation the loss or destruction of forests, mainly for logging or farming

delta alluvial deposit at a river's mouth that looks like the Greek letter delta

democracy any system of government in which leaders rule with consent of citizens

demographic transition a model that uses birth rates and death rates to chart changes in the population trends of a country or region

demography the study of population

deregulate to remove restrictions and regulations

desalination the removal of salt from seawater to make it usable for drinking and farming

desertification the process in which arable land is turned into desert

developed country a country that has a great deal of technology and manufacturing

developing country a country in the process of becoming industrialized

dharma in Hinduism, a person's moral duty, based on class distinctions, which guides his or her life

dialect a local form of a language used in a particular place or by a certain group

dike a large bank of earth and stone that holds back water

dissident a citizen who speaks out against government policies

divide a high point or ridge that determines the direction rivers flow

doldrums a frequently windless area near the equator

domestic relating to, made in, or done in one's own country

domesticate to adapt plants and animals from the wild to make them useful to people

dominion a partially self-governing country with close ties to another country

doubling time the number of years it takes a population to double in size

dry farming a farming method used in dry regions so that land is plowed and planted deeply to hold water in the soil

dynasty a ruling house or continuing family of rulers, especially in China

dzong a fortified monastery of Bhutan, South Asia

Eastern Hemisphere the part of Earth east of the Atlantic Ocean including Europe, Asia, Australia, and Africa; longitudes 20° W and 160° E often considered its boundaries

e-commerce selling and buying on the Internet

economic of, relating to, or based on the production, distribution, and consumption of goods and services

economic sanction a trade restriction

ecotourism the practice and business of recreational travel based on concern for the environment

elevation the height above the level of the sea

El Nino-Southern Oscillation (ENSO) a seasonal weather event that can cause droughts in Australia and powerful cyclones in the South Pacific

emir a prince or ruler in Islamic countries

enclave a region or community, as within a country or city, made up of people of a different race or cultural background

endemic native to or belonging to a particular environment

Enlightenment a movement during the 1700s that emphasized the importance of reason and questioned traditions and values

environmentalist a person actively concerned with the quality and protection of the environment

equator the parallel of 0 degrees latitude from which other latitudes are calculated

equinox one of two days (about March 21 and September 23) on which the sun is directly above the equator, making day and night equal in length

erosion the wearing away of the earth's surface by wind, flowing water, or glaciers

escarpment a steep cliff or slope between a higher and lower land surface

estuary an area where the tide meets a river current

ethnic of or relating to races or large groups of people classed according to common traits and customs

ethnic cleansing the expelling from a country or killing of rival ethnic groups

ethnic group a group of people who share common ancestry, language, religion, customs, or a combination of such characteristics

European Union (EU) an organization whose goal is to unite Europe so that goods, services, and workers can move freely among member countries

eutrophication the process by which a body of water becomes too rich in dissolved nutrients that stimulate aquatic plant growth, usually resulting in depleted oxygen

exclave a distinct group of people who are isolated from the main or larger part of a country

export a commodity sent from one country to another for purposes of trade

extended family a household made up of several generations of family members

extinction the disappearance or end of an animal or plant species

farm cooperative an organization in which farmers share in growing and selling farm products

fault a crack or break in the earth's crust

fauna the animal life of a region

federal system a form of government in which powers are divided between the national government and the state or provincial government

feudalism in medieval Europe and Japan, a system of government in which powerful lords gave land to nobles in return for pledges of loyalty

fisheries areas, freshwater or saltwater, in which fish or sea animals are caught

fjord a long, steep-sided glacial valley now filled by seawater

flora the plant life of a region

flow-line map a map that shows the movement of people, animals, goods, ideas, and physical processes like hurricanes and glacial movement

foehn a dry wind that blows from the leeward sides of mountains, sometimes melting snow and causing avalanches; term used mainly in Europe

fold a bend in layers of rock, sometimes caused by plate movement

formal region a region defined by a common characteristic, such as production of a product

fossil fuel a resource formed in the earth by plant and animal remains

free port a port city, such as Singapore, where goods can be unloaded, stored, and reshipped without the payment of import duties

free trade the removal of trade barriers so that goods can flow freely between countries

free trade zone an area of a country in which trade restrictions do not apply

functional region a central place and the surrounding territory linked to it

genetically modified food food sources, the genes of which have been altered to cause increase in size and speed of growth or greater resistance to pests

geographic information systems computer tools for processing and organizing details and satellite images with other pieces of information

glaciation the process whereby glaciers form and spread

glacier a large body of ice that moves across the surface of the earth

glasnost the Russian term for a new openness in areas of politics, social issues, and media; part of Mikhail Gorbachev's reform plans

global economy the merging of resource management systems so countries are interconnected and dependent on one another for goods and services

globalization the increasing interconnectedness of people and places throughout the world through converging processes of economic, political, and cultural change

global warming the gradual warming of Earth and its atmosphere that may be caused in part by pollution and an increase in the greenhouse effect

globe a spherical representation of Earth

glyphs picture writing carved in stone

graziers farmers who raise sheep or cattle in New Zealand

great circle route an imaginary line that follows the curve of Earth and represents the shortest distance between two points

Greater Antilles the four large Caribbean islands of Cuba, Jamaica, Hispaniola, and Puerto Rico

Great Escarpment this landform rims southern Africa from Angola to South Africa. It forms where the narrow coastal plains meet the elevated plateaus in an abrupt break in elevation

greenhouse effect the capacity of certain gases in the atmosphere to trap heat, thereby warming Earth

green revolution the program begun in the 1960s to produce higher-yielding, more productive strains of wheat, rice, and other food crops

gross domestic product (GDP) the value of goods and services created within a country in a year

gross national product (GNP) is similar to GDP but is a broader measure including the inflow of money from other countries in the form of the repatriation of profits and other returns on investments, as well as the outflow to other countries for the same purposes

groundwater water within the earth that supplies wells and springs

guest worker a foreign laborer working temporarily in an industrialized, usually European, country

guru a teacher or spiritual guide

habitat an area with conditions suitable for certain plants or animals to live

haiku a form of Japanese poetry originally consisting of 17 syllables and three lines, often about nature

hajj in Islam, the yearly pilgrimage to Makkah (Mecca)

harmattan a dust-laden wind that blows on the Atlantic coast of Africa in some seasons

headwater the source of a stream or river

heavy industry the manufacture of machinery and equipment needed for factories and mines

hemisphere half of a sphere or globe, as in Earth's Northern and Southern Hemispheres

hierarchical of, relating to, or arranged in order of rank

hieroglyphic belonging to an ancient Egyptian writing system in which pictures and symbols represent words or sounds

Holocaust the mass killing of 6 million Jews by Germany's Nazi leaders during World War II

homogeneous of the same or similar kind or nature

horticulture the science and art of growing fruits, vegetables, and plants

human-environment interaction the study of the interrelationship between people and their physical environment

hurricane a large, powerful windstorm that forms over warm ocean waters

hydroelectric power electrical energy generated by falling water

hydrosphere the watery areas of the earth, including oceans, lakes, rivers, and other bodies of water

hypothesis a scientific explanation for an event

Ibadhism a conservative form of Islam distinct from the Sunni and Shia sects

ideogram a pictorial character or symbol that represents a specific meaning or idea

ideology ideas, characteristics of a person, group, or political party

immigrant a person who comes to a country to take up permanent residence

imperialism the actions by which one nation is able to control other, usually smaller or weaker, nations

indentured laborers foreign workers contracted to labor on Caribbean agricultural estates for a set period of time, often several years. Usually the contract stipulated paying off the travel debt incurred by the laborers. Similar indentured labor arrangements have existed in most world regions.

industrialization the transition from an agricultural society to one based on industry or manufacturing

infrastructure the basic urban necessities like streets and utilities

insular constituting an island, as in Java

introduced species plants and animals placed in areas other than their native habitat

Inuit a member of the Arctic native peoples of North America

Iron Curtain a term coined by British leader Winston Churchill during the Cold War that defined the western border of Soviet power in Europe. The notorious Berlin Wall was a concrete manifestation of the Iron Curtain

Islamic fundamentalism A movement within both the Shiite and Sunni Muslim traditions to return to a more conservative, religious-based society and state. Often associated with a rejection of Western culture and with a political aim to merge civic and religious authority

Jainism A religious group in South Asia that emerged as a protest against orthodox Hinduism about the sixth century BC. The ethical core of Jainism is the doctrine of noninjury to all living creatures

Japan Current a warm-water ocean current that adds moisture to the winter monsoons

jati in traditional Hindu society, a social group that defines a family's occupation and social standing

jazz a musical form that developed in the United States in the early 1900s blending African rhythms and European harmonies

jute the plant fiber used to make string and cloth

karma in Hindu belief, the sum of good and bad actions in one's present and past lives

key a map legend or guide

Khmer Rouge Literally, "Red (or communist) Cambodians." The left-wing insurgent group led by French-educated Marxists who rebelled against the royal Cambodian government, first in the early 1960s and then again in a peasants revolt in 1967

kibbutz a collective farm in Israel

kolkhoz in the Soviet Union, a small farm owned by a collective that paid farmers as salaried employees

kum the term for a desert in Central Asia

lagoon a shallow pool of water at the center of an atoll

lama the Buddhist religious leader

landlocked enclosed or nearly enclosed by land

language family a group of related languages that have all developed from one earlier language

latifundia in Latin America, large agricultural estates owned by families or corporations

latitude the distance north or south from the equator measured in degrees

leach to wash nutrients out of the soil

leeward being in or facing the direction toward which the wind is blowing

Lesser Antilles the arc of small Caribbean islands from St. Maarten to Trinidad

Levant the eastern Mediterranean region

light industry manufacturing aimed at making consumer goods, such as textiles, or food processing rather than heavy machinery

lingua franca a common language used among people with different native languages

literacy rate the percentage of people in a given place who can read and write

lithosphere the surface land areas of the earth's crust, including continents and ocean basins

llano the fertile grassland in inland areas of Colombia and Venezuela

lode a deposit of minerals

loess fine, yellowish-brown topsoil made up of particles of silt and clay usually carried by the wind

longitude the distance measured by degrees or time east or west from the prime meridian

Loyalist a colonist who remained loyal to the British government during the American Revolution

Maastricht the site of a 1992 meeting of European governments in the Netherlands, at which a treaty was signed forming the European Union

Maghreb a region in northwestern Africa, including portions of Morocco, Algeria, and Tunisia

magma molten rock that is pushed up from Earth's mantle

maharaja regional Hindu royalty, usually a king or prince, who ruled specific areas of South Asia before independence, but who was usually subject to overrule by British colonial advisors

mantle the thick middle layer of Earth's interior structure, consisting of dense, hot rock

mantra in Hinduism, a sacred word or phrase repeated in prayers and chants

manuka a small shrub that grows in plateau regions of New Zealand

Maori the indigenous Polynesian people of New Zealand

map projection a mathematical formula used to represent the curved surface of Earth on the flat surface of a map

maquiladora in Mexico, a manufacturing plant set up by a foreign firm

marianismo an idealized woman stressing the virtues of patience, deference, and work in the home

marine west coast climates moderate climates with cool summers and mild winters heavily influenced by maritime conditions. It is usually found on the west coasts of continents between latitudes of 45 and 50 degrees. Climates are moderated by their proximity to oceans or large seas. They are usually cool, cloudy, wet, and lack the temperature extremes of continental climates.

maritime concerned with travel or shipping by sea

market economy an economic system based on free enterprise; businesses are privately owned, and production and prices are determined by supply and demand

marsupials mammals whose offspring mature in a pouch on the mother's abdomen

martial law the control and policing of civilians by military rules

matriarchal describes a woman who rules her family and descendants

megacity a city with more than 10 million people

megalopolis a thickly populated area centered around several large and small cities, or one large city

Melanesia the Pacific Ocean region that includes the culturally complex, generally darker-skinned peoples of New Guinea, the Solomon Islands, Vanuatu, New Caledonia, and Fiji

meltwater water formed by melting snow and ice

merchant marine a country's fleet of ships that engage in commerce or trade

mestizo refers to biracial people of Native American and European descent

metropolitan area the region that includes a central city and its surrounding suburbs

Micronesia the Pacific Ocean region that includes the culturally diverse, generally small islands north of Melanesia. Includes the Mariana Islands, Marshall Islands, and Federated States of Micronesia.

Middle Ages the period of European history from about A.D. 500 to about A.D. 1500

migration the movement of people from place to place

minifundia in Latin America, small farms that produce food chiefly for family use

mistral a strong northerly wind from the Alps that can bring cold air to southern France

mixed economy a system of resource management in which the government supports and regulates enterprise through decisions that affect the marketplace

mixed farming raising several kinds of crops and livestock on the same farm

mixed forest a forest with both coniferous and deciduous trees

monarchy a form of autocracy with a hereditary king or queen exercising supreme power

monopoly the total control of a type of industry by one person or one company

monotheism belief in one God

Monroe Doctrine a proclamation issued by U.S. President James Monroe in 1823 that the United States would not tolerate European military action in the Western Hemisphere. Focused on the Caribbean as a strategic area, the doctrine was repeatedly invoked to justify U.S. political and military intervention in the region.

monsoon in Asia, seasonal wind that brings warm, moist air from the oceans in summer and cold, dry air from inland in winter

moraine piles of rocky debris left by melting glaciers

mosque in Islam, a house of public worship

mujahedeen Islamic guerrilla fighters

mural a wall painting

nationalism the belief in the right of each people to be an independent nation

natural boundary a fixed limit or extent defined along physical geographic features like mountains and rivers

natural increase the growth rate of a population

natural resources substances from the earth that are not man-made

natural vegetation plant life indigenous to an area

newly industrialized country a country that has begun a transition from primarily agricultural to primarily manufacturing and industrial activity

nomad a member of a wandering pastoral people

North American Free Trade Agreement (NAFTA) the trade pact made in 1994 by Canada, the United States, and Mexico

nuclear of, relating to, or using the atomic nucleus, atomic energy, the atom bomb, or atomic power

nuclear family a family group made up of a husband, wife, and children

nuclear proliferation the spreading development of nuclear arms

nuclear waste the by-product of nuclear power production

oasis a small area in a desert where water and vegetation are found

oligarchy a system of government in which a small group holds power

oral tradition stories passed down from generation to generation by word of mouth

organic farming the use of natural substances rather than chemical fertilizers and pesticides to enrich the soil and grow crops

Organization of African Unity (OAU) founded in 1963, the organization grew to include all the states of the continent except South Africa, which finally was asked to join in 1994. It is mostly a political body that has tried to resolve regional conflicts.

Organization of American States (OAS) founded in 1948 and headquartered in Washington, D.C., the OAS advocates hemispheric cooperation and dialogue. Most states in the Americas belong, except Cuba.

outsourcing the practice of subcontracting manufacturing work to outside companies, especially foreign or nonunion companies

overfishing harvesting fish to the extent that certain species are depleted and the fishing area made less valuable

ozone layer the atmospheric layer with protective gases that prevents solar rays from reaching Earth's surface

pampas grassy, treeless plains of southern South America

Parliament in British-influenced areas, the national legislature that is made up of the Senate and the House of Commons

pastoralism the raising of livestock

patois dialects that blend elements of indigenous, European, African, and Asian languages

patriarchal relating to a social group headed by a male family member

perceptual region a region defined by popular feelings and images rather than by objective data

perestroika in Russian, "restructuring;" part of Gorbachev's plan for reforming Soviet economy and government

permafrost a permanently frozen layer of soil beneath the surface of the ground

pesticide a chemical used to kill insects, rodents, and other pests

petrochemical a chemical product derived from petroleum or natural gas

phosphate a natural mineral containing chemical compounds often used in fertilizers

physical map a map that shows the location of natural features such as mountains and rivers; can also show cities and countries

pidgin simplified speech used among people who speak different languages

pidgin English a dialect mixing English and a local language

planar projection a map created by projecting an image of Earth onto a plane

plate tectonics the term scientists use to describe the activities of continental drift and magma flow, which create many of Earth's physical features

poaching illegal hunting of protected animals

pogrom in czarist Russia, an attack on Jews carried out by government troops or officials

polder a low-lying area from which seawater has been drained to create new farmland

political map a map that shows the boundaries and locations of political units such as countries, states, counties, cities, and towns

pollution the existence of impure, unclean, or poisonous substances in the air, water, or land

Polynesia the Pacific Ocean region, broadly unified by language and cultural traditions, that includes the Hawaiian Islands, Marquesas Islands, the Society Islands, the Tuamotu Archipelago, Cook Islands, American Samoa, Samoa, Tonga, and Kiribati

polytheism belief in many gods

population density the average number of people per square mile or square kilometer

population distribution the pattern of population in a country, a continent, or the world

postindustrial refers to an economy with less emphasis on heavy industry and manufacturing and more emphasis on services and technology

prairie an inland grassland area

precipitation moisture that falls to the Earth as rain, sleet, hail, or snow

primate city a city that dominates a country's economy, culture, and government and in which population is concentrated; usually the capital

Prime Meridian the meridian of 0 degrees longitude from which other longitudes are calculated

privatization a change to private ownership of state-owned companies and industries

prophet a person believed to be a messenger from God

province an administrative district or division of a country

puna a treeless, windswept tableland or basin in the higher Andes

radioactive material material contaminated by residue from the generation of nuclear energy

rain shadow effect the result of a process by which dry areas develop on the leeward sides of mountain ranges

raj the Hindi word for "empire"

realism an artistic style portraying everyday life that developed in Europe during the mid-1800s

reforestation planting young trees or seeds on lands where trees have been cut or destroyed

refugee one who flees his or her home for safety

regime a form of government

region a broad geographical area distinguished by similar features

relief the elevations or inequalities of a land surface

Renaissance in Europe, a 300-year period of renewed interest in classical learning and the arts, beginning in the 1300s

reparation a payment for damages

restriction a regulation that confines, limits, or restrains

retooling converting old factories for use in new industries

revenue the income produced by a given source

revolution in astronomy, the Earth's yearly trip around the sun, taking 365 days

rice paddy flooded field in which rice is grown

Richter scale measures the amount of energy released at the epicenter of an earthquake

rift valley a crack in Earth's surface created by shifting tectonic plates

Romanticism the artistic style emphasizing individual emotions that developed in Europe in the late 1700s and early 1800s as a reaction to industrialization

samurai in medieval Japan, a class of professional soldiers who lived by a strict code of personal honor and loyalty to a noble

satellite a country controlled by another country, notably Eastern European countries controlled by the Soviet Union by the end of World War II; also a form of telecommunications involving the beaming of radio signals into the atmosphere so they can be reflected back again

savanna a tropical grassland containing scattered trees

scale the size of a picture, plan, or model of a thing compared to the size of the thing itself

scale bar a map key that shows the relationship between map measurements and actual distance on Earth

sect a subdivision within a religion that has its own distinctive beliefs and/or practices

sedentary farming farming carried on at permanent settlements

separatism the breaking away of one part of a country to create a separate, independent country

serf laborer obliged to remain on the land where he or she worked in feudal times

service industry a business that provides a service instead of manufacturing goods

shantytown a makeshift community on the edge of a city

shari'ah Islamic law derived from the Koran and the teachings of Muhammad

sheikhdom a territory ruled by an Islamic religious leader

shields large upland areas of very old exposed rocks that range in elevation from 600 to 5,000 feet (200 to 1,500 meters)

shifting cultivation clearing forests to plant fields for a few years and then abandoning them

shogun a military ruler in medieval Japan

sickle a large, curved knife with a handle, used to cut grass or tall grains

sirocco hot desert wind that can blow air and dust from North Africa to Western Europe's Mediterranean coast

situation refers to the geographic position of a place in relation to other places or features of a larger region

slash-and-burn farming a traditional farming method in which all trees and plants in an area are cut and burned to add nutrients to the soil

smog haze caused by the interaction of ultraviolet solar radiation with chemical fumes from automobile exhausts and other pollution sources

socialism the political philosophy in which the government owns the means of production

socialist realism the realistic style of art and literature that glorified Soviet ideals and goals

solstice one of two days (about June 21 and December 22) on which the sun's rays strike directly on the Tropic of Cancer or Tropic of Capricorn, marking the beginning of summer or winter

sovereignty self-rule

Soviet era the period between 1921 and 1991 when Russia was part of the Union of Soviet Socialist Republics

sovkhoz in the Soviet Union, a large farm owned and run by the state

spreading a process by which new land is created when sea plates pull apart

station an Australian term for an outlying ranch or large farm

steppe the wide, dry plains of Eurasia; also, similar semiarid climate regions elsewhere

Strine colloquial English spoken in Australia

stupa a dome-shaped structure that serves as a Buddhist shrine

subcontinent a large landmass that is part of a continent but still distinct from it, such as India

subduction a process by which mountains can form as sea plates dive beneath continental plates

subsidy a grant or gift, especially of money

subsistence agriculture farming that produces only enough crops or animal products to support a farm family's needs. Usually little is sold at local or regional markets.

suburb an outlying community around a city

Sunbelt the mild climate region in the southern United States

supercell a violent thunderstorm that can spawn tornadoes

sustainable development technological and economic growth that does not deplete the human and natural resources of a given area

syncretism a blending of beliefs and practices from different religions into one faith

taiga the Russian term for the vast subarctic forest, mostly evergreens, that covers much of Russia and Siberia

tariff a tax on imports or exports

technology the use of science in solving problems

tectonic plate a massive, irregularly shaped slab of Earth's crust and uppermost mantle, composed of both continental and oceanic lithosphere

temperature the degree of heat or cold measured on a set scale, such as Fahrenheit or Celsius

thematic map a map that emphasizes a single idea or a particular kind of information about an area

tierra calienta the Spanish term for "hot land"; the lowest altitude zone of Latin American highlands climates

tierra fria the Spanish term for "cold land"; the highest altitude zone of Latin American highlands climates

tierra helada the Spanish term for "frozen land"; a zone of permanent snow and ice on the peaks of the Andes

tierra templada the Spanish term for "temperate land"; the middle altitude zone of Latin American highlands climates

timberline the elevation above which it is too cold for trees to grow

topography the shape of Earth's physical features

tornadoes twisting funnels of air with winds up to 300 mph

total fertility rate the average number of children a woman has in her lifetime

trade deficit spending more money on imports than earning from exports

trade surplus earning more money from export sales than spending for imports

tsunami a very large sea wave induced by underwater earthquakes

tundra the arctic region with a short growing season in which vegetation is limited to low shrubs, grasses, and flowering herbs

typhoons large tropical storms, similar to hurricanes that form in the western Pacific Ocean in tropical latitudes and cause widespread damage to the Philippines and coastal Southeast and East Asia

unitary system a government in which all key powers are given to the national or central government

universal suffrage equal voting rights for all adult citizens of a nation

urbanization the movement of people from rural areas into cities

urban sprawl the spreading of urban developments on undeveloped land near a city

viceroy a representative of the Spanish monarch appointed to enforce laws in colonial Latin America

wadi a streambed that is dry except during a heavy rain in the desert

wat a temple in Southeast Asia

water cycle the regular movement of water from ocean to air to ground and back to the ocean

wattle a woven framework made from acacia saplings by early Australian settlers to build homes

weather the condition of the atmosphere in one place during a short period of time

weathering chemical or physical processes, such as freezing, that break down rocks

welfare state a nation in which the government assumes major responsibility for people's well-being in areas such as health and education

Western Hemisphere the half of the Earth comprising North and South America and surrounding waters; longitudes 20° W and 160° E often considered its boundaries

windward being in or facing the direction from which the wind is blowing

World Trade Organization (WTO) an international body that oversees trade agreements and settles trade disputes among countries

NOTES

NOTES

DISCARD

910 GEOGRAPHY

Geography review in
20 minutes a day

METRO

R4001305720

METROPOLITAN
Atlanta-Fulton Public Library